The Impact of the Internet

ISSUES

Volume 31

Editor

Craig Donnellan

Educational Publishers
Cambridge

First published by Independence
PO Box 295
Cambridge CB1 3XP
England

British Library Cataloguing in Publication Data
The Impact of the Internet – (Issues Series)
I. Donnellan, Craig II. Series
004.6'78

ISBN 1 86168 215 8

Printed in Great Britain
MWL Print Group Ltd

Typeset by
Claire Boyd

Cover
The illustration on the front cover is by Pumpkin House.

CONTENTS

Chapter One: The Internet Society

Chapter Two: Safety on the Internet

Introduction

The Impact of the Internet is the thirty-first volume in the ***Issues*** series. The aim of this series is to offer up-to-date information about important issues in our world.

The Impact of the Internet examines the use of the Internet in society and the issue of safety on the Internet.

The information comes from a wide variety of sources and includes:
Government reports and statistics
Newspaper reports and features
Magazine articles and surveys
Literature from lobby groups
and charitable organisations.

It is hoped that, as you read about the many aspects of the issues explored in this book, you will critically evaluate the information presented. It is important that you decide whether you are being presented with facts or opinions. Does the writer give a biased or an unbiased report? If an opinion is being expressed, do you agree with the writer?

The Impact of the Internet offers a useful starting-point for those who need convenient access to information about the many issues involved. However, it is only a starting-point. At the back of the book is a list of organisations which you may want to contact for further information.

*****

Internet access

Information from National Statistics

Households and individuals

Over the period October to December 2001 an estimated 9.8 million households in the UK could access the Internet from home, according to the Expenditure and Food Survey (EFS). That amounts to almost four in 10 (39 per cent) of all UK households.

Fifty-six per cent of adults in Great Britain have accessed the Internet at some time according to figures from the February 2002 National Statistics Omnibus Survey. This is equivalent to 25.6 million adults in Britain having accessed the Internet. In the month prior to the survey 46 per cent of adults had accessed the Internet.

Households with home access to the Internet

Over the fourth quarter of 2001 an estimated 9.8 million households in the UK could access the Internet from home representing 39 per cent of all households. This is the same proportion as found in the last quarter (July to September 2001) but is over four times the number three years earlier. Over 2001 the growth in household Internet access has begun to level off from the steady increases seen in the previous two years.

Figures from April 2000 cover all forms of access including new technologies such as digital TV; earlier figures only include access from home computers. In the October to December 2001 quarter only a small percentage of households (two per cent) accessed the Internet exclusively using technologies other than home computers.

Individual access to the Internet

Fifty-six per cent of adults in Britain have accessed the Internet at some time according to figures from the February 2002 National Statistics Omnibus Survey. This is equivalent to 25.6 million adults in Britain having accessed the Internet. This is a rise of five percentage points from the figure in January 2001. Over 80 per cent of individuals who had accessed the Internet did so in the month prior to the survey (46 per cent of the adult population).

As found in July and October 2001, the gap between men and women who have used the Internet has closed considerably. While the proportion of men accessing the Internet has remained relatively stable between January 2001 (57 per cent) and February 2002 (58 per cent), the proportion of women accessing the Internet has increased by nine percentage points from 45 per cent to 54 per cent over this period.

The proportion of adults who had used the Internet decreases steadily with age from over 80 per cent of those aged 16 to 24 years to 12 per cent of those aged 65 and over.

Uptake of Internet access

The Internet is continuing to attract new users, with 10 per cent of those who have ever accessed the Internet, doing so for the first time within the last three months.

Within the last year more women users (29 per cent) than men users (20 per cent) reported logging on to the Internet for the first time. Nearly half (49 per cent) of all men who had accessed the Internet reported having logged on over three years ago, compared with just over a third (36 per cent) of women.

How individuals access the Internet

The focus of the National Statistics Omnibus Survey is to collect information on individuals' personal use of the Internet. Six per cent of those adults who had accessed the Internet had done so exclusively for work and are therefore excluded from the subsequent figures presented here.

Individuals can now use a range of technologies to access the Internet. Computers continue to dominate as the preferred method for accessing the Internet. Ninety-nine per cent of individuals who used the Internet for personal use had done so using a computer. By February 2002 nine per cent of adults who had ever used the Internet had done so using a mobile phone. As found in previous quarters around six per cent reported using Digital Television.

Where people use the Internet

While individuals continue to access the Internet for personal use from a wide range of locations, the respondent's own home continues to be the most popular location (79 per cent) followed by their workplace (35 per cent) or another person's home (28 per cent).

Frequency of access

Respondents were also asked how frequently they used the Internet. Men still use the Internet more than women: 62 per cent of men who accessed the Internet for private use did so more than once a week (30 per cent at least once a day), while 48 per cent of women who had accessed the Internet for private use, did so more than once a week (23 per cent at least once a day). Conversely 24 per cent of women who accessed the Internet for personal use did so less than once a month; the equivalent figure for men was 14 per cent.

Respondents were also asked when they had last accessed the Internet for personal use. Sixty-nine per cent reported accessing the Internet within the last seven days and a further 13 per cent within the last month; just three per cent had last accessed the Internet over a year ago.

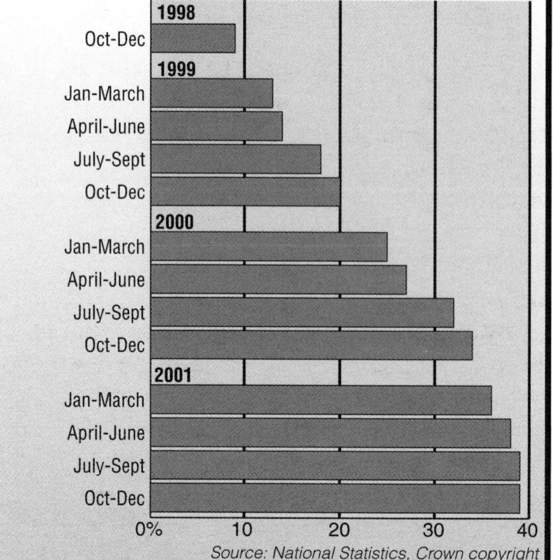

What people do on the Internet

Among those who had accessed the Internet for private use nearly three-quarters of adults used the Internet to find information about goods or services (74 per cent) or used e-mail (73 per cent), while over half used it for general browsing (56 per cent). Since January 2001 the proportion of adults ordering tickets, goods or services has risen by 12 percentage points to 42 per cent.

Internet purchases

Respondents who had already stated that they used the Internet for private use and who bought or ordered tickets, goods and services or used the Internet for personal banking, financial and investment activities were asked what goods they had bought on the Internet in the three months prior to interview. The most popular purchases were flights and holiday accommodation (28 per cent), books or magazines (27 per cent), music or CDs (25 per cent) and tickets for events (21 per cent). A further 20 per cent of respondents who had previously bought goods on the Internet had not bought anything in the three months prior to interview. Of those who had purchased goods or services in the last three months, 42 per cent reported spending £100 or less in the three months prior to interview, while 10 per cent spent over £500. Respondents' main reasons for never having purchased over the internet was that they had security concerns (23 per cent), or that they preferred to shop in person (21 per cent). Although nearly a quarter of adults stated that they had security concerns only four per cent of adults who had used the Internet for private use stated that they had actually experienced any security problems.

People who do not use the Internet

Those adults who had never accessed the Internet (44 per cent) were asked why they had not used it. A wide range of reasons was given and respondents could give more than one answer if they wished. Forty-three per cent of those who had not used the Internet stated that they were not interested in using it, 25 per cent had no means of access to the Internet, 21 per cent did not feel that they had the confidence or the skills required to use the Internet while a further 17 per cent felt they had no need to access the Internet.

Just over half (52 per cent) of all individuals who have yet to access the Internet reflected a general lack of interest in accessing the Internet[1] when reporting their main reasons for not accessing the Internet.

In February 2002 72 per cent of adults who had never accessed the Internet stated that they were very unlikely to access the Internet in the next year. This core group represents nearly a third (32 per cent) of all adults.

Note

1 No interest in the Internet, no need to use the Internet and do not want to use the Internet.

• The above information is an extract from a release by National Statistics. For more information see their web site: www. statistics.gov.uk

59% of Britons now using interactive technologies

Egg/MORI report sees around 1 million new users of online banking since October – now at c. 6.6 million users

'Technology is increasingly helping to enrich our lives. The internet, e-mail and text message are all parts of today's vocabulary and as consumer confidence grows with transacting online, the world-wide web will become an integral part of everyday life for all.'

Patrick Muir, Marketing Director, Egg UK

The latest Egg/MORI 'Embracing Technology' Report, issued today, reveals Britons continue to demonstrate a voracious appetite for technologies such as the internet, e-mail and instant messaging. But more than that, the nation displays strong faith in technology making a real difference to their lives, whether through the invention of cars running on a renewable, cheap resource or being able to make payments through mobile phones.

Key Highlights:

- The Egg Index: 59% of the British adult population – or around 27 million people – now use interactive technologies (internet, digital TV or WAP mobile phone) for personal use (October 2001: 54%).
- We are confirming our status as a nation of digital bankers: approximately 6.6 million Britons now bank online (October 2001: approx 5.5 million).
- Click happy: Internet use continues to rise fast with 45% of the British public – or around 20 million – now using the internet for personal use (October 2001: 42%). A further 1 in 10 envisage accessing it within the next two years, indicating sound growth potential for online financial services.
- Cheque's in the dustbin: A third of all adults (31%) say they would be interested in being able to make payments to individuals or pay bills to businesses using e-mail, as a faster and more efficient alternative to posting cheques. This figure grows to a half (54%) among e-mail users.
- Account aggregation: Over a third of all adults (37%) say they would be interested in account aggregation facilities in the next 2 years.
- Online financial servicing: Half of internet users (around 10 million adults) say they have either arranged or serviced a financial product over the internet.

The Egg Index – measuring the change in interactive technology use

Between October 2001 and February 2002, the Egg Index has recorded an increase of around 2 million British adults using interactive technologies (internet, digital TV or a WAP mobile phone) for personal use. There are now approximately 27 million adults using either the internet, digital TV or WAP. Internet penetration has reached 45% of British adults, up from 42% in October, an increase of around 1 million new users.

For the first time, this report examines the uptake of instant messaging on the internet. Instant messaging allows internet users to have virtual conversations online in real time. This is claimed to be used by around 6 million people, or 14% of adults in Britain, already. Uptake grows to 52% amongst experienced internet users – those that have been online for 5 years or more – and 31% among all internet users, indicating this online tool is one to watch.

A gaping gulf in technology uptake between Cardiff and Croydon

Amidst the continued uptake of these technologies, there are some strong regional differences in terms of usage levels. For instance, 56% of those living in the South East (excluding London), and 50% of Londoners, use the internet – compared to only 23% in Wales. A similar pattern is evident for e-mail usage. While 16% of all British adults claim to use laptop PCs, this rises to one-quarter among Londoners, and falls to just 5% in Wales.

A nation of online bankers

Continuing to grow in harmony with the growth of the internet universe is the number of people banking online. The Egg Report has seen a growth of around 1.1 million people doing this in the four months since October 2001 alone. This compares to a growth of around 1.5 million between April and October last year, as shown in our previous report.

Uptake of online financial products
The findings show that half of internet users say they have now either bought or serviced a financial product over the internet. This amounts to around 10 million British adults. Looking forward, there is a huge growth potential for online financial services; 52% of internet users say they would consider arranging some sort of financial product over the internet in the next two years.

Aggregating our finances online
With almost half (45%) of British adults now online, the internet has truly penetrated our lives, and we have reached a stage where we are hungry to conduct a greater part of our day-to-day lives online. The most popular future financial tool to be surveyed amongst our representative sample of British adults is account aggregation. Over a third of all adults (37%) – some 17 million people – say they would be interested in being able to view and access all the financial products they hold with all their providers, using a single internet site and one log-in, in the next 2 years.

Amongst internet users this figure rises to an even higher proportion – 61% – and to three-quarters (75%) among those who currently use online banking.

Patrick Muir, Marketing Director, Egg UK, observes: 'The advent of online banking did away with queuing at branches and allowed people to service their finances any time of the day or night. This report reveals consumers expect the internet to continue to improve their lives even further, with the introduction of account aggregation services that allow them to manage their whole financial lives online.'

Cheque's in the post? No thanks – we want speed payments
The humble cheque may die a death at the hands of modern technology. Being developed today is a faster, more efficient payment requiring only the touch of a button. Over half (54%) of e-mail users – and 31% of all adults – say they would be interested in the ability to make payments to individuals or pay bills to businesses using e-mail in the next two years.

Braving the battle of the bulge
Carrying wallets laden with heavy change, or being caught without a penny, are annoyances that one in five of us may want to do away with. The Egg Report shows 19% of all adults would be interested in making payments from mobile phones using SMS text messaging in the next two years, rising to 34% amongst those who currently use text messaging.

• MORI conducted a total of 1,965 interviews with a nationally representative sample of adults aged 16+ across Great Britain. All interviews were conducted face to face, in home between 7th and 13th February 2002. The data were weighted to reflect the national profile.

Where figures do not sum to 100 per cent this may be due to computer rounding, multiple codes or exclusion of 'Don't Know'.

Throughout this report, comparisons are made with the last Egg report – 'Embracing Technology' published December 2001. A total of 1,959 interviews were conducted face to face, in home between 18th and 24th October 2001.

Online shopping in Britain increases

In February 2002, the number of people in Britain shopping online was almost 20% higher than 12 months before. Consequently, online spend grew by almost 20% to £4.3bn in the past six months.

Average total spend did not rise over the period, remaining at about £470, but the average spend on a single purchase did climb, to £164 in February, compared with £134 a year earlier.

This reflects the variety of items being bought. As well as books and CDs, travel, clothes, electrical goods and even cars are rising in popularity.

Those who have bought online in the past six months say the most important factor in their choice of site is security, followed by trust in the company, ease of use and price. More than half say each of these is 'very important'. Only 37% say time of delivery is 'very important'.

However, satisfaction with online shopping is declining. In February 2001, 68% said their last experience was 'very good', but this year the proportion was 61%. This may be due to rising expectations caused by consumer familiarity with the web.

Online shopping summary

Number of internet users in past month
17.3 million

Web users purchasing online in the past six months
9.2 million

Average per head spend for past six months
£466

Total value of online shopping in past six months
£4.3 billion

• The above information is from BMRB International's web site which can be found at www.bmrb.co.uk

Computers in schools

One million computers in schools but is everybody headed for an online future?

The latest research from the British Educational Suppliers' Association (BESA) suggests that while the teaching profession is taking on board the £1.8 billion government revolution in ICT, it is still struggling to give pupils access to an online future.

Thanks to increases in government spending, the current stock of computers in schools has grown 24% in the past year to reach the one million mark, although 25% of these are still classified as 'ineffective for curriculum use'. In 2000 there were an estimated 27 desktop computers per school – now there are 34. Growth is expected to continue in 2002 with 70% of schools indicating new purchases by April 2002. Schools indicate that the optimum number of desktop computers required to implement current ICT development plans is 1.4 million. Once new growth in 2002 is included schools will have 1.1 million computers. If this expansion continues at the current rate, schools will meet their current needs in 2005.

There is no dispute that almost 100% of schools now have an Internet connection. As a result of this it is estimated that 75% of computers in school have Internet access. This equates to 26 computers per school; a significant leap from the 1999 figure of 12 per school. However, 70% of schools are connected using ISDN as their networked access route to the Internet. The average bandwidth across all schools limits use to an average of six concurrent users (4.5 in primary and 14.7 in secondary) accessing interactive websites with significant streaming. Such a factor must make us think realistically about an immediate online future for schools.

The new BESA research shows a significant swing, especially in the primary sector, to locate computers out of the classroom in computer 'labs'. Overall, 56% of schools have made a conscious decision to move computers into a separate designated area. Only special schools keep the majority of their machines in the immediate teaching environment. This new statistic must bring into question the hope that computers will 'pervade every lesson' when timetabling restrictions are taken into account. This new way of managing teaching and learning has been shown to favour the teaching of discrete ICT rather than a cross-curricular focus. It must also narrow the access to an online solution, rather than extend it.

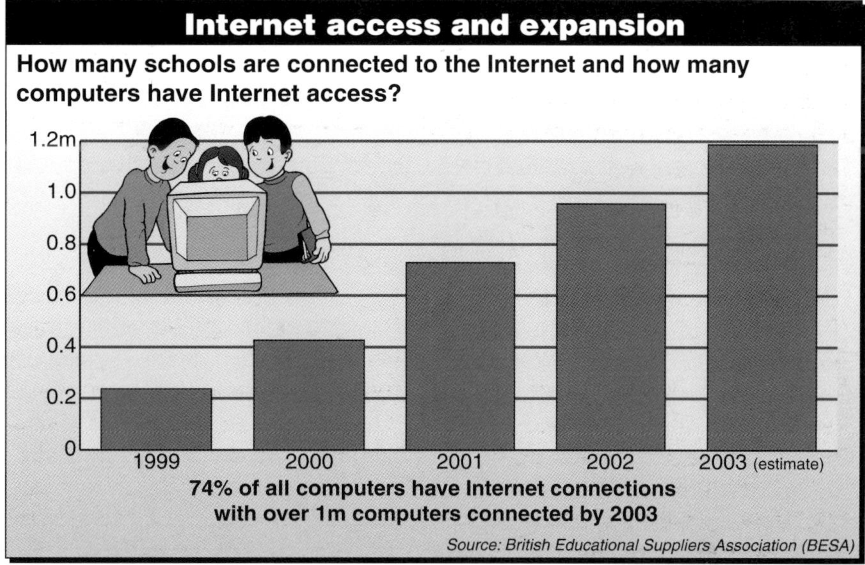

There is no dispute that almost 100% of schools now have an Internet connection. As a result of this it is estimated that 75% of computers in school have Internet access

With practically all schools having some form of connection to the Internet the focus in schools has moved to accessibility for teachers and pupils. This can be seen in the dramatic rise in the number of laptops in schools. There has been a significant expansion in the past year from 55,000 units to 100,000 units. The number of schools owning at least one laptop has grown from 62% in 2000 to 87% in 2001. There is now an average of 4.4 laptops per school. The drive for connectivity has also led schools to consider alternative more innovative technologies. Wireless networking is gaining interest with 35% of schools considering implementation to allow the mobility of networked laptops.

Dominic Savage, Director General of BESA, said, 'It is clear that teachers have moved forward ICT in schools over the past year with over 60% feeling confident and competent. However, the figures in the BESA survey suggest that we must not presume that everything is going to be carried out online at this time and this must influence government policy concerning the Curriculum Online initiative. If this real revolution is to be successful in schools we must be assured of continued government funding for hardware, content and training.'

- The above information is from BESA's web site which can be found at www.besanet.org.uk

The net is not the place for learning

Websites provide an overwhelming amount of information – and much is nonsense, says Simon Webb

The internet is rapidly eclipsing the encyclopaedia as the first place a pupil turns to in search of basic information. Is this a trend that should concern us? Surely the medium in which knowledge is stored is of little importance? Isn't the crucial point that it should be accurate and easily accessible?

Yet is precisely on these grounds that the internet, as it is currently being used in schools, fails miserably.

For the average child, retrieving information with a computer is a time-consuming and bewildering process.

Compare 'looking it up in a book' and 'searching the net' and see how the two methods work in practice. My 11-year-old daughter wanted to find out about slow worms. The *Encyclopaedia Britannica* in the local library provided the necessary information in less than five minutes.

The *Britannica* is also available on the internet, but children seldom begin their search by trying to remember the address of its website (www.britannica.com).

Instead, they call up a search engine, type the key words into a box and then click 'search'. It is a procedure that generates an unbelievably long list of websites, most of which will turn out to be quite irrelevant.

In the case of slow worms, more than 400 references were produced by my daughter's enquiry on a terminal at her school.

Wasting time is undesirable, but the really worrying aspect of the internet is that young people have absolutely no way of judging the quality of a site

If there are three screen pages or so per site, this gives us around 1,200 pages to work through – roughly the length of *War and Peace*.

As a literate adult, I have the skills to enable me to skim through this huge mass of information and extract those parts that might be of use. Even so, it is a daunting task.

Children tend either to click on each site in turn and plough through them all or simply copy out the first thing they come to. The first method wastes an enormous amount of time, the second is statistically unlikely to yield anything of value.

For instance, the slow worms search turned up hundreds of sites with no connection whatever with slow worms.

Some of them would prove irresistibly attractive to a bored child. Several were about a computer game called Worms, one of which had colourful animations. It is hard to imagine many 11-year-olds so dedicated to the task in hand that they would not be tempted to waste time on such distractions,

Wasting time is undesirable, but the really worrying aspect of the internet is that young people have absolutely no way of judging the quality of a site.

Many concerns have been expressed about the possibility of a child coming across pornography on

the internet but hardly anyone seems to be alarmed at the quantity of misleading and inaccurate rubbish to which they will inevitably be exposed.

Recently, a friend of my daughter surprised her teacher by asserting, in a history project, that American Indians had arrived in the country 2,000 years ago in a fleet of Phoenician ships. The explanation was both simple and disturbing. While researching on the net, she had stumbled across the information on a website operated by the Mormon Church.

How can we possibly expect children of this age to weigh the rival merits of a website run by the Mormons against one run by the Smithsonian Institute in Washington, DC? The problem is that anybody in the world can set up a website on any subject and then post anything they wish upon it. This is quite different from the traditional means of disseminating knowledge or opinion: no peer review, no critical publishers, no competing for space in newspapers or magazines.

Had I written a book 30 years ago claiming that the Earth was flat, it is unlikely that any publisher would have touched it. Even if one did, it is inconceivable that I could have ensured that every school and library in the country acquired a copy. This, in effect, is what the internet allows me to do.

I can set up a website tomorrow propounding my unorthodox cosmology, and every pupil researching a homework project will then be liable to encounter it.

How are they to be expected to distinguish between the sensible, the outlandish and the frankly mad?

For proof of this, one has only to type the word 'evolution' into a search engine to discover the huge number of creationist websites.

In the past, great care has been taken in selecting books for schools and children's libraries. Now, with an increasing number of schools giving out no text books, but blithely advising their pupils to 'look it up on the net', the dangers are clear.

As a matter of interest, our search discovered no authoritative site about slow worms. The most accurate information came from a website run by a woman in the West Country who obsessively records every detail of her garden and then makes the data available on the internet. She found a slow worm in the garden and decided to feature it on her site.

I could not resist e-mailing her to ask where she had learnt so much about slow worms. She replied that she had looked them up in an encyclopaedia.

Computer games 'help pupils learn'

By Macer Hall and Peter Warren

Schools are to use computer games to help pupils to learn under a government scheme designed to raise standards by bringing fun into the classroom.

The football game Championship Manager, the motorsport simulation F1 Championship Racing and the city-building scenario Sim City will be used to prepare teaching materials for lessons.

The decision follows a pilot project carried out by the British Educational Communications and Technology Agency (Becta), a research organisation.

The project, which was backed by the Department for Education and Skills, attempted to show that computer games could help children to absorb information and acquire useful skills.

Last night, however, there was concern that the idea would lead to children wasting hours of school time. Chris Woodhead, the former chief inspector of schools and now an outspoken campaigner for traditional teaching methods, condemned the move as 'cobblers'.

Mr Woodhead said: 'Huge sums of money are being wasted. There are simply not enough teachers who are competent enough to use technology or competent to evaluate the value of what is being taught at the moment.'

A spokesman for the DfES said: 'The department is interested in exploring the potential of computer games as a tool to aid learning and teaching.'

Research showed that the games could develop skills such as problem-solving and memorisation as well as helping children to learn from each other, the spokesman added.

The Internet and education

Information from Becta (British Educational Communications and Technology Agency)

What is the Internet?

The Internet is a huge network of computers making a worldwide community, with millions of members, providing a vast store of information with great possibilities for education. More than 80% of all primary schools, all colleges and virtually all secondary schools already have an Internet connection and the Government's target is for all schools, colleges and libraries to be connected by 2002. The home-user market is also expanding swiftly.

The Internet offers a range of facilities, allowing users to obtain information and resources, to communicate with each other and to publish information. Some of the most frequently used facilities include:

The World Wide Web or Web provides easy access to the vast quantity of information and resources available on the Internet and is the facility which people use to 'surf' for information. It is made up of millions of screens, or 'pages', of information. The collection of pages created by one individual or organisation is known as a website. Each page can include text, images, sound, animation and video and has its own unique address.

E-mail allows users to send and receive written messages via a telephone line. Students have used e-mail for communicating with pen pals, to send questions to a specialist (such as a vulcanologist) to help with project work, and to swap information, for example, about their locality and weather with students in other countries.

Mailing lists usually consist of a group of people who exchange e-mail about a subject that interests them. For instance, a mailing list called 'ukschools' allows people with an interest in education to write to the list about any relevant topic, alert one another to useful websites or ask for help and advice.

Internet Service Providers (ISPs) provide the link between the user and the Internet. Some provide a free service and with others, charges vary according to the type of connection, amount of use and additional services such as filtering.

Newsgroups are like international noticeboards where people log in to a particular group to read and contribute remarks or questions. There are thousands of newsgroups covering any and every topic and interest, including some which could be considered offensive. Consequently, some ISPs choose not to offer their subscribers all groups, or do not carry news at all. Nonetheless, older students have benefited from access to newsgroups, which have helped to expand their knowledge and allowed them to discuss specific topics in depth.

Chat rooms allow a number of people to 'meet' on the Internet and have live, 'real-time' conversations. It is similar to having a telephone conversation with a number of people at once except that the participants type instead of talk. Open chat lines are not often used in schools, to avoid any possible compromise to pupil safety. However, chat rooms are very popular with pupils who have on-line access at home. Individual websites, such as those produced by the BBC (http://www.bbc.co.uk/education/schools/) and the Science Museum (http://www.nmsi.ac.uk/) are increasingly setting up specific times for question-and-answer sessions with subject experts or well-known personalities, which have relevance to the school curriculum. Childnet International, the children's Internet charity, have launched a new website to respond to the growing dangers for children using Internet chat rooms called www.chatdanger.com.

Video conferencing enables two or more people, in different locations, to see one another while they talk. Secondary schools, in particular, have used video conferences as a resource for language learning, for example. It is also possible to arrange for the exchange of audio, video, images or any other digital file to allow users in different places to work concurrently on the same resource.

Any individual or organisation can create and publish a website. Most ISPs offer free Web space to their subscribers.

Educational benefits of the Internet

Information and communications technology (ICT) projects run by the UK Education Departments have shown that the use of ICT in education provides a number of learning benefits, including:

- improved subject learning across a wide range of curriculum areas, including English, maths, science, history, geography, modern languages, art, technology, IT and careers, as well as independent study and cross-curricular project work
- improved motivation and attitudes to learning
- development of independent learning and research skills
- improved vocational training
- development of network literacy (i.e. the capacity to use electronic networks to access resources, create resources and communicate with others – these can be seen as complex extensions of the traditional skills of reading, writing, speaking and listening)
- social development.

Through the Internet, students are able to 'visit' places and take part in events that would otherwise be too far away, or too costly. For example, the 24 Hour Museum enables visitors to take a virtual tour, the Hubble Space Telescope offers full-colour pictures of planets, while the Whole Brain Atlas gives access to colour MRI scan images of the entire brain. Other sites are specifically designed to support education in the UK, such as the National Grid for Learning (NGfL) which offers students, teachers and parents access to a growing collection of curriculum-related resources and educational support material, offering access to lifelong learning, basic skills and community information.

Learners have always needed to be able to find relevant and reliable information quickly and easily, and to select, interpret and evaluate what they find. Searching for information on the Web can help to develop these information skills. For example, the Web is an ideal way to introduce and illustrate concepts such as audience and purpose, authorship and bias. By viewing sites with opposing views, the idea of subjectivity/objectivity, fact/opinion and overt/covert bias can be demonstrated in an 'immediate' and powerful way.

While educators and parents need to exercise caution when allowing children access to the Internet, they should not be deterred from using it

What should teachers and parents be aware of?

The Internet is a reflection of the people who make up our society. It is not controlled by any particular organisation and the standard or source of information cannot always be guaranteed. Individuals' rights to freedom of speech and freedom of choice must be observed, but balanced against the rights of younger users. Although not common, users will also wish to guard against the possibility of hacking and computer viruses.

While educators and parents need to exercise caution when allowing children access to the Internet, they should not be deterred from using it. Its educational benefits outweigh any possible dangers. Schools have always helped learners to engage with society, based on clear support and guidance, and use of the Internet should be no exception.

As with television and video, parents, carers and educators should preview material or provide supervision, as well as having a more general strategy in place for ensuring children's safe use of the Internet.

These strategies can use a combination of measures. Different circumstances call for different approaches so that young children can be protected from exposure to inappropriate material while older students may need no more than simple logging and monitoring. Some schools and colleges use an open service: they may supervise learners or make them earn their access. Some use an Internet service that filters material, while others also buy control or filtering software to monitor computer activity.

One of the best ways for parents to become comfortable with their children's Internet access is to become actively involved. By learning to use the Internet alongside their children, parents can instil the values that they want their children to use when selecting material, not just on the Internet, but on television, radio, or in print.

A *sensible approach involves:*

- siting computers in public places where everyone can see what is on the screen
- taking an interest in the Internet and regularly discussing what young people see and use
- being aware of what research projects children are carrying out on the Internet
- monitoring on-line time and being aware of excessive hours spent on the Internet
- educating students to use the Internet in a sensible and responsible manner
- encouraging learners to be critical users of the Internet: 'Is the information true? How do you know?'
- warning young people that there are some unsuitable sites on the Internet and discussing the issues involved
- warning young people that people may try to contact them in an inappropriate way and discussing the issues involved
- setting up a reporting system so that students know what to do if they find upsetting material
- requiring all learners to sign a code of conduct and, for those under 18, sending copies to parents for them to sign

- making it clear what the penalties are for misuse of the Internet
- ensuring that learners are aware that they must not respond to abusive messages.

Website evaluation techniques can also have an important role to play. Children should be encouraged to ask questions about a site, such as:

- Authority

Who has written the information? What is the authority or expertise of the author? Are there contact details for the author? Who is the publisher of the site?

- Purpose

What are the aims of the site? Does it achieve its aims?

- Audience

Who is the site aimed at?

- Relevance

Is the site relevant to me?

- Objectivity

Is the information offered as fact or opinion? Is the information overtly or covertly biased?

- Accuracy

Can the accuracy of the information be checked?

- Currency

When was the information written? When was the site last updated?

- Format

Does it contain information in the format that I want?

- Links

Does the site give me advice/ideas/ other choices?

Rural area have-nots lose out on the net

Technology underclass must be given right of access, warns report

By Stuart Millar, Technology Correspondent

The internet revolution has created a new underclass of people in rural and remote areas who are being excluded from the brave new world of teleworking, virtual shopping and online public services by lack of access to technology.

That is the conclusion of research, to be published this week, which warns the much-hyped potential for new technologies to 'render distance obsolete' is not being realised.

Despite confident predictions that the expansion of information and communications technologies (ICT) would bridge social divides – between urban and rural and rich and poor – the UK's knowledge economy currently threatens to bypass large areas of the country, leaving them 'trapped in a cycle of low skills/low value work'.

The research, carried out by the Local Futures Group thinktank for IT giant IBM, calls on the government to treat access to technology as just as important as access to transport and health care.

'Those unable to travel easily to work, to access ICT, shop or socialise are in danger of becoming excluded from our increasingly mobile society, missing out on the opportunities available, and penalised for their lack of mobility,' the authors conclude. Rural communities stand, in theory, to gain most from technologies which allow them to work, shop or access services electronically.

But the researchers found that home-based teleworking remains an urban phenomenon, rising fastest in London where congestion and the costs of commuting provide good reasons for people to want to work from home.

The image of someone using new technology to work from their remote Highland cottage remains largely a myth.

Part of the problem, according to the authors, has been failure to identify demand and social requirements. The failure of teleworking in the countryside, for example, may be because working life still requires regular face-to-face contact with employers, colleagues and clients.

'Too much research has focused on new technologies and markets, on the possibilities rather than the realities of human communication, and ignored the social context of people's needs, desires and patterns of behaviour, both at work and leisure.'

The report concludes that mobile technologies offer the best platform for breaking this cycle, with services from customised weather reports and skills training to remote health care and e-democracy all delivered through wireless networks to mobile phones, notebook computers and handheld devices. Mobility is described as 'the killer ICT application in rural areas'.

One of the most promising technologies is interactive digital television, which experts believe has a better chance of penetrating rural areas than conventional PC-based internet connections. Already embryonic services such as interactive advertising and e-mail are proving extremely popular.

But achieving change will require more effective leadership and intervention from the public sector, the report warns. The weak consumer base and dispersed nature of business has meant that rural areas have been overlooked in the market-led roll-out of broadband communications services, a situation which will not be reversed unless public sector bodies such as local councils, schools and health trusts take the lead.

The researchers also criticise the government for 'missing a trick' by failing to find innovative ways of using mobile technologies to bridge the digital divide between urban and

rural. They point out that while it is possible to get bus timetables via a mobile in London, this service is unheard of in the vast majority of rural areas.

Paul Revell, IBM's wireless e-businesses service executive, said: 'The big challenge with rural areas is to ensure that they can join the virtual community but that is going to take government working hand in hand with the private sector.'

Kate Oakley, the group's director of research projects and co-author of the report, said: 'Immobility is an aspect of exclusion. ICT should help break down that problem because it gives virtual mobility but . . . the take-up has just followed normal socio-economic lines so existing inequalities have just been reinforced.'

She added: 'If you accept that ICT access should be part of the infrastructure then you can't leave it to the market because it will take too long and the gap will get wider.'

UK passes 500,000 high-speed Internet connections

The UK now has over half a million high-speed broadband connections, Oftel announced today.

Broadband services are now delivered to homes and businesses using cable modems, DSL technology, broadband fixed wireless and broadband satellite services.

Welcoming the news, David Edmonds, Director General of Telecommunications, said:

'Over 500,000 customers now have high-speed Internet access using a broadband connection, according to current published figures from the network operators.

'The 500,000 doesn't take into account the cable operators' connections for March and April, so the actual figure is significantly more.

'With over 20,000 broadband connections a week, the current level of growth outstrips the equivalent demand for mobile phones and dial-up Internet when they were first introduced.

'And the UK has more competition at both network and service levels than many European countries.

'Where many consumers in Europe rely on the incumbent for broadband services, UK consumers are choosing from a number of different networks and service providers.

'Over 10 million homes use the traditional dial-up Internet access, including four million with unmetered packages.

'I am confident that more Internet users will take up high-speed broadband as the range of services increases and prices fall.'

David Edmonds said that Oftel had taken action to create competition at all levels of the broadband market.

'The UK now has some of the cheapest Internet prices in the world. As a result, the number of people signing up to broadband is accelerating'

'Competition has resulted in growing take-up rates and falling prices.

'Oftel has promoted competition between different networks and different service providers to ensure consumers get a wide range of choice and prices.

'New technology such as satellite, fixed wireless and 3G mobile phones means there are an increasing number of networks competing to deliver broadband services to consumers.'

Welcoming the milestone of half a million broadband connections, E-Commerce Minister Douglas Alexander said:

'The UK now has some of the cheapest Internet prices in the world – for both narrowband and broadband. As a result, the number of people signing up to broadband is accelerating.

'The milestone of half a million connections represents a 54 per cent increase since the beginning of 2002. Of course there is more to do, but the work of building Broadband Britain is under way.

'BT's commitment to broadband and their recent price reductions, together with the successful cable broadband programmes of NTL and Telewest, pave the way for further advances on broadband during 2002.'

The widening digital divide

The latest internet use survey reveals that despite initiatives to bring more people online, a gulf in skills, opportunity and interest remains, writes Neil McIntosh

You might not have read about it in today's newspapers, but a significant new internet use study was published yesterday by the UK office for national statistics. The muted reception for its findings is in stark contrast to that afforded its predecessors.

Granted, through the tech mania of the late 1990s, internet surveys made amazing reading. The rapid rate of internet adoption was outstripping any technology that had gone before – even television and radio – sending shockwaves through the media industry and helping create the crazy valuations and predictions that were a signature of the times.

Now the bubble has burst, those valuations have been spectacularly deflated and the predictions are of doom, rather than glory. Last autumn one newspaper reported the popularity of the internet was 'crashing', while others suggested millions of users were logging off.

Neither was correct. The new study shows 'newbies' are still arriving in cyberspace – and, indeed, there has been a modest spike in uptake in recent months.

Between April 2001 and March 2002, 40% of UK households were connected. But the first three months of this year found 42% of households connected, a 6% increase year on year, suggesting that a steady rate of increase will continue for a while at least.

One high-street giant is certainly banking on it. Last week Dixons announced it was looking to recruit 1000 new staff to accommodate what it predicts will be a boom in consumer electronics. New products like wireless home networks and a new generation of personal digital assistants will lead the way, but the company even sees new life in the moribund personal computer market too.

Meanwhile sales of fast broadband internet connections are warming up, in part thanks to lower prices and more aggressive marketing from British Telecom. Take-up so far suggests broadband will take longer to establish itself as a widespread means of connecting to the internet – we're talking millions signing up over years, rather than months – but the situation has improved dramatically from only a year ago.

> ***80% of the richest bracket of households have internet access, against 11% of the poorest***

So is all rosy in the internet garden? Not quite. Read deeper into the figures and they suggest that, while there may be continued short-term growth in household internet use, it is likely to hit a plateau soon.

Dixons might be getting ready for a rush, but step inside one of its shops and you quickly discover that even the cheapest means of hooking up to the net are pretty expensive. Even a comparatively cheap £600 PC is beyond the means of many – a further £300 a year minimum for broadband, plus charges for installation and equipment, equally out of reach.

There is clear evidence of a widening digital divide: adoption remains far higher among the young and the well off. Almost half of households in London and the south-east of England have net access, against 31% in Wales and Northern Ireland, 34% in the West Midlands and 35% in the south-west.

80% of the richest bracket of households have internet access, against 11% of the poorest.

Among non-internet users, there remains a hardcore of people who say they are unlikely to use the internet any time soon. In the latest survey, 72% of non-users say they are very unlikely to access the internet in the next year. This group represents nearly a third of all adults.

It would be easy to brand these refusniks as Luddites. 44% of them say they haven't got an interest in using the internet. But 25% say they lack a computer, or access to one, and 20% say they lack confidence or the skills to try it out.

While those who have never used the internet might be surprised by the breadth of its content they are not going to get a chance to find out if access is denied to them in the first place.

Technology will catch up with some. Digital TV will offer more and more a form of internet access that will feel more like teletext than the web. Today's mobile phones, which enjoy near-saturation levels of ownership in the UK, will be replaced by devices with net-style information services and e-mail fitted as standard in the years to come.

But, despite government initiatives to bring internet access and, more recently, broadband, to less densely populated parts of the country, it remains clear a gulf in skills, opportunity and interest remains. Britain's first 40% of internet penetration has been the easy bit: the next 25% could be a far tougher nut to crack.

Who's got mail?

At work, e-mail and the Web become public. By Andrew Bibby

In a few short years, e-mail and the Internet have transformed business practice in many countries. Companies have found that they have powerful new tools available to them, both for communicating and for obtaining information. But what happens when being able to 'surf while you work' becomes more interesting than the job at hand? Journalist Andrew Bibby examines the benefits – and drawbacks – of this workplace revolution

In today's office, this sentence could have gone around the world faster than you could write it.

So-called 'snail mail' – communications that used to move as fast as the planes, boats and trains could carry them – have been replaced by e-mail that can circle the globe in moments rather than days or weeks.

Faxes are becoming like postage – old, slow and virtually ignored.

And all employees are 'on-line', that is, have a relatively high-speed Internet connection at their work stations.

Welcome to the electronic office. It's fast, it's efficient, and it's fraught with potential problems. Granted, it's made communications a breeze. Yet the implications are as vast as the area covered.

E-mail rules

The speed of change has left employers, workers' representatives and indeed individual workers struggling to know what sort of rules should govern the way these technologies are used at work. Is it acceptable, for example, for an employee to surf the internet at work for purposes which are not directly work-related? Is there a difference between visiting a web site, say, on health and safety issues and simply catching up on the latest sports results?

And to what extent should employers be able to monitor their employees' use of e-mail and the Internet? Is it acceptable, for example, for an employer to trawl through the e-mails sent by an individual worker to the trade union? Does it make a difference if employer monitoring is publicised rather than being covert?

In the early days of the Internet, a certain laissez-faire approach to questions like these may have sufficed. In fact, some employers actively encouraged workers to become familiar with the technology, taking the view that staff who surfed the Web for fun would end up better trained and more productive. Other companies adopted an informal approach simply because they hadn't worked out a coherent management stance.

This is increasingly less of an option. For example, it is becoming clear that, in some circumstances, companies can be called to account for e-mails sent by their employees. Perhaps the highest profile example of the use of company e-mails was the US anti-trust hearing against Microsoft, when the US Justice Department made use of private messages sent by Microsoft chairman Bill Gates, as part of its case against the company.

There have been other occasions when the texts of e-mails have become public knowledge – and companies have rued the fact that they were ever sent. One example was that of a US oil giant which lost a sexual harassment case and paid out $2.2m in damages to four women workers when internal e-mails were shown to reveal an aggressively male work-culture. In the United Kingdom, a large insurance company was obliged to pay £450,000 ($650,000) to a rival insurer after its staff were found to have sent libellous e-mail messages about the other company. Another concern for companies is that e-mail addresses used at work usually include the company name, suggesting that e-mail messages are in some sense official communications. This was a serious problem for one Swedish firm, when an employee used a work e-mail address to send messages of support to an extreme right-wing organisation.

E-mails are also a conduit for importing computer viruses to internal corporate networks. The Melissa virus alone has been estimated to have cost North American businesses $80m. The 'iloveyou' virus went around the world in hours, infecting many large and small companies' systems.

Monitoring mail and web

It is perhaps unsurprising, therefore, if employers are increasingly introducing policies on e-mail and Internet use and are engaging in monitoring the use which employees make of these technologies. The best evidence of this trend comes from the US, where the American Management Association (AMA) has been surveying this issue each year for the past four years. The AMA reports that the percentage of major US companies who store and review employees' e-mail messages increased from 27% in 1999 to 38% in 2000. As recently as 1997, only 15% of firms engaged in this practice.

It is a similar story with Internet use. According to the AMA, an even larger number of US companies, about 54% in 2000, are now monitoring the Internet connections made by their employees.

Coupled with these developments has been an increase in disciplinary action taken against employees who are deemed to have breached the rules. For example, the *New York Times* sacked 23 of its staff in November 1999 for e-mailing jokes and pornographic pictures. The following summer, an international investment bank sacked fifteen of its London employees who were also alleged to have circulated offensive material by e-mail.

But moves like these can be controversial, particularly in the broader context of industrial relations in the workplace. The airline company Ansett sacked one of its employees for what it called an 'unacceptable use of technology'. The woman, a delegate of the Australian Services Union, had circulated to colleagues via e-mail an update on the bargaining talks being carried on between the company and the union. The case was referred up to the Australian Federal Court, who found in April 2000 in the employee's favour. The airline, the court found, had contravened freedom of association provisions in the country's Workplace Relations Act.

A new workplace issue

Trade unions have been engaging in the issues raised by cases such as these for a number of years. One of the first initiatives was that taken by the international trade union federation FIET (now part of Union Network International, UNI), which launched a campaign for on-line rights for on-line workers early in 1998. UNI has taken over the running of the campaign, and co-hosted an international conference on the key themes in Brussels at the end of last year.

UNI's campaign identifies a number of separate, but interrelated, issues. Firstly it claims that there is an issue of freedom of association to be addressed. UNI argues that, in an increasingly electronic world of work, workers' organisations should have access as of right to electronic means of communication to reach members and potential members, to engage in the normal process of industrial relations. As UNI's General Secretary Philip Jennings puts it, 'Trade unions appreciate the advantages which new communication technologies can bring, and know that in the electronic workplace the old ways of communicating with employees may no longer be the most appropriate.'

Welcome to the electronic office. It's fast, it's efficient, and it's fraught with potential problems

The provision of facilities to workers' representatives was an issue which the ILO addressed in 1971 in both Convention No. 135 and the accompanying Recommendation. The Recommendation talks of trade unions having the right, for example, to post notices and to distribute news-sheets and publications to workers. UNI argues that these rights should extend to electronic forms of communication, particularly as new working methods such as teleworking take an increasing number of people away from the centralised workplaces of the past. One of UNI's on-line rights campaign demands is for trade unions, works councils and individual workers to have access to corporate e-mail systems for industrial relations purposes. A second demand is for employees to have the right of free access to trade union web sites, and other Internet sites relevant to their rights at work.

These issues have been picked up by a number of trade union organisations in individual countries. In South Africa, for example, the trade union federation Cosatu adopted a declaration in August 1999 which committed the organisation 'to specifically launch a campaign to ensure dedicated access for each shop steward to a computer, internet and e-mail facilities at each workplace'. A similar demand is now included in the Charter of Workplace Union Delegates' Rights, adopted by the Australian Council of Trade Unions.

But UNI has also raised a more fundamental issue, the extent to which electronic surveillance and monitoring ('snooping', according to its critics) represents an unacceptable invasion of an individual's right to privacy. Legal expert Professor Gillian Morris told a recent conference that whilst employers' legitimate interests might mean some encroachment, in her opinion there should remain an 'inviolable zone of privacy' for employees which employers should not intrude within.

Finding a solution

A number of countries are currently attempting to reconcile these issues, in the context of privacy laws. In the Netherlands, the Chamber of Registration (the Dutch body charged with privacy legislation) published advice in January which would allow employers the right to check employee e-mails and internet usage, provided that clear ground-rules have first been drawn up and made public. The German Ministry of Labour has announced its intention to proceed with an 'employee data protection act'. In the UK, a draft Code of Practice being drawn up by the country's Information Commissioner has been criticized by employers for being too tightly drawn.

Issues of access to electronic facilities and of electronic monitoring are also being tackled within the context of normal industrial relations. A number of unions, including GPA in Austria, FNV in the

Netherlands and MSF in the UK, have produced model agreements covering e-mail and Internet use. The FNV model, for example, proposes that employees should have the right to use e-mail and the Internet for non-commercial purposes 'provided that this does not interfere with their day-to-day work commitments' (the right to deliberately visit pornographic or racist websites is specifically excluded, however). In France the multimedia union Betor-Pub CFDT has negotiated a formal agreement with the Société OLSY, which gives the trade unions the right to use internal electronic means of communication to keep in touch with members.

A recent report from the French Commission nationale informatique et libertés (CNIL) suggests that a common-sense approach can help establish fair and effective e-mail and Internet usage policies in companies. CNIL proposes a relatively relaxed approach to employee use of e-mail and the Web, for instance, with private use permitted within reasonable limits, provided that it does not prevent the normal work use of these channels of communication. Banning all private e-mails would be an unrealistic step, it suggests.

CNIL says that companies should develop clear and detailed

policies on security and monitoring, which should be made public to employees. Analysis of Internet usage specifically by individual workers should not normally take place, expect in 'exceptional circumstances'. CNIL adds that employers might legitimately forbid visits to certain types of web site, such as pornographic or Holocaust-denial sites.

Despite these developments, the next few years could see even greater difficulties in establishing good practice in this area of employment relations and employment law. Technology is changing: employees will increasingly be able to send e-mail and to search Internet sites whilst away from their desks by using their mobile phones whilst at the same time their phones will potentially allow their employers at all times to monitor their exact geographical location. At the same time, new ways of working are developing, which are seeing the formerly rigid boundaries between work and personal life become more and more blurred. Fair and sensible procedures for employee use of electronic forms of communications seem likely to become even more essential in the years ahead.

• Andrew Bibby is a journalist specialising in on-line issues.

• The above information is an extract from *World of Work* – a magazine produced by the International Labour Organization.

Laugh? I nearly got the sack

John Crook lost his job for sending a lewd joke by e-mail. The incidence is rising, warns Fran Abrams, as companies crack down on 'inappropriate' humour

First, a word of warning. If you ever receive an e-mailed dinner invitation that purports to come from your boss, do not send a reply that reads: 'I have made plans to spend this evening driving nails through my toes – therefore I am far too busy to consider going anywhere with you.' No. The smart thing to do is to take a good, hard, sideways look at the office prankster. If he is doubled up in giggles, delete the e-mail and say nothing. Believe me, I know. It's easy to be wise after the event.

Most office jokes are harmless, of course. There has been no reported case yet of an employee sacked for running an electronic sweepstake on what time the laziest man in the building will log on to the computer system; or of anyone losing their job for passing on one of those joke lists of workplace terminology containing gems such as: 'What's a consultant? Answer: any ordinary bloke more than 50 miles from home.'

But many employers – particularly the large corporations – are taking an increasingly tough stance on office humour that may be construed as sexist, racist or just plain rude.

Last week, John Crook, a regional manager for the recruitment firm Manpower Services, had cause to rue the day he cracked a joke in an e-mailed memo to his boss, Angela Brunton. Recommending a colleague for a pay rise, Crook had praised her work rate and then added: 'And she was a grrrreat shag as well.' Crook, who worked in Norwich, was sacked.

Last Thursday, an industrial tribunal turned down his claim for unfair dismissal, rejecting Crook's explanation that he knew Brunton well and believed that she would find the remark amusing. The colleague about whom he was writing had seen the e-mail and had also thought it funny, he added, but his explanation that 'it was in the context of a culture I'd become used to' fell on deaf ears.

John Crook's name is thus added to a growing list of those who have come to grief after taking office humour just that bit too far.

There have been other cases. Late last year, Cable and Wireless dismissed six staff for sending smutty e-mails, and a Huddersfield-based turbo-charge maker, Holset Engineering, successfully defended the sacking of two employees for a similar offence.

Perhaps the best-known case was that of Bradley Chait, a lawyer with the London firm Norton Rose. He was disciplined last December after passing on a flattering message from a girlfriend, Claire Swire, about his sexual prowess. His boastful e-mail was subsequently bounced around the globe, first by colleagues and later by complete strangers, accompanied by the challenge: 'Where is Claire Swire?'

Chait's case demonstrates the havoc a single e-mail can cause. Smutty office jokes have always been passed around, of course, but in the past they were circulated on scraps of paper, copied and recopied so many times that they became almost illegible. Now, with one click of a button, a lone message can be projected into dozens of different workplaces and from there can multiply with almost unimaginable speed.

Earlier this year, no fewer than 10 employees at Royal & Sun Alliance insurance were sacked and a further 77 suspended after the circulation of an e-mail that showed the cartoon character Bart Simpson flashing at his naked sister Lisa. A subsequent investigation was reported to have unearthed a number of other such visual jokes, including one showing Bart's mother Marge performing oral sex on her clean-living neighbour, Ned Flanders. Another showed Kermit the Frog in a compromising pose with Fozzie Bear.

Many large companies are now making it clear to their workers that such jokes are not considered acceptable.

Steve Field, the UK head of employee services for KPMG business consultants, says he would consider the passing on of e-mails showing cartoon characters in lewd situations a sacking matter.

'The advice that jumps to my mind is that you should never, ever send anything by e-mail that you wouldn't show across a crowded room'

'I would certainly dismiss for that kind of offence,' he says. 'If it was initiated or passed on by our employee, and if it then went outside our organisation, it would be badged as being from KPMG, and that's exposing us to risk.' Field remembers the days before e-mail and says material of 'dubious character' was certainly passed around then, too. But now, KPMG deals with such incidents at the rate of about two a month – a far higher level than before. 'The amount and the nature of the inappropriate material has changed,' he says. 'Originally you might have got jokes, but now you tend to get visual messages – even animations.'

Sue Sadler, a spokeswoman for Marks & Spencer, says the company now tells workers they will be breaking their terms of employment if they pass on jokes that may be seen as malicious, defamatory or offensive. 'We have had cause to discipline people in the past because of this,' she says.

She adds, though, that standards of acceptable behaviour are stricter for those who work on the shop floor than for those in offices. While it may be acceptable to crack an inoffensive joke to a colleague at the next desk, giggling with a workmate while customers wait unattended is considered the worst possible level of service.

Some firms still take a more relaxed attitude. Jason Fisher, UK managing director of eCircle, a company that runs e-mail operations, says he could hardly afford the time to monitor the communications of its 100 or so employees.

'At the end of the day you don't have control,' he says. 'Even if we wanted to be rather fascist about it, we don't have the resources. The guys in the office sometimes forward a couple of jokes, but it's never been an issue for me. As long as they're doing their jobs, it's fine – you have to trust people.'

Diane Sinclair, employee relations adviser with the Chartered Institute of Personnel and Development, says it is important for all companies to make sure that their workers understand what is acceptable and what is not.

'The advice that jumps to my mind is that you should never, ever send anything by e-mail that you wouldn't show across a crowded room,' she says. 'I think people feel that e-mail is private, that it's only coming onto one person's screen – but that isn't the case.'

Although office culture has grown to be more informal in many ways, with most employees and their bosses on first-name terms, legislation now outlaws jokes that can be construed as sexual harassment or discrimination. 'Obviously it's a positive thing if people have good relations and enjoy their workplace,' says Sinclair. 'But it has to be the right kind of humour in the right environment.'

In some places, however, office humour is considered not just desirable – but obligatory. Some firms in India now start the day with a 'laughter club' at which senior staff are abandoned to the mercy of a man with a megaphone and a joke book.

Maybe it should catch on here, too. A hearty dose of workplace chuckles might even put the office prankster out of business.

• This article first appeared in *The Guardian*, 5 June 2001.

Post modern

By Oliver Burkeman

It was billed as instant, democratic and cheap, and it was going to revolutionise the way we communicated. But less than a decade after it became widely available, e-mail is losing its lustre. Many of us are complaining of information overload, and companies are grappling with a proliferation of pointless communication.

The revolution started on a Friday. In a greyish building on a sprawling business park outside Watford, a maverick executive at the offices of Camelot, the national lottery operator, issued a startling edict: no more e-mails on the last day of the working week unless totally, absolutely necessary. No pointless cc-ing of irrelevant memos; no links to 'comedy' websites intended to amuse; no electronic invitations to the pub after work. Staff, it seemed, were forgetting how to talk to each other.

'We needed to make staff more aware of other forms of communication,' said a Camelot spokeswoman. 'If there were elements of the business where you could talk face to face instead of sending an e-mail, we wanted to encourage people to do that.'

Four weeks later, the experiment came to an end. The company is 'still in the process of reviewing the results', but it is hard to imagine that the initiative will prove to have been anything but a triumphantly futile gesture. According to a survey published this week, we are at risk of drowning in e-mail, struggling to keep our heads above the virtual water as 6.1bn electronic messages slosh daily around the planet – and, as a result, we are turning against it as never before.

The Consumers' Association found that the number of people who named e-mail as their favourite form of communication had plummeted, in the space of a year, from 14% to 5%. Meanwhile, the numbers who preferred face-to-face meetings jumped from 39% to 67%. Old-fashioned 'snail mail', other reports show, is holding up well, too – diminishing as a percentage of all communications, but growing in volume by 2% a year worldwide. Across the globe, technological sceptics – the cockles of their hearts already warmed by the crash in internet stocks – are smiling smiles of vindication.

The Consumers' Association found that the number of people who named e-mail as their favourite form of communication had plummeted, in the space of a year, from 14% to 5%

We used to be able to dismiss the spectre of information overload as a fantasy of panicky luddites. In 1998 – when office workers were beginning to hyperventilate at receiving a mere 20 or 30 e-mails a day – a typical Mori survey of the British working population noted, in a tone of calm reassurance, that 'e-mail volume is by no means as heavy as many scaremongers lead people to believe'. But that was three years ago, when the idea of connecting to the internet on a train through a laptop connected to a cellphone, or through Blackberrys – the nifty, pocket-sized e-mail-checker that combines keyboard, screen and wireless dialer, now de rigueur in what remains of Silicon Valley – was barely credible to many.

Now, plans are afoot to facilitate e-mail-checking via cellphones in the tunnels of the New York subway, and London Underground has promised – threatened? – the same. Mori undertook a similar survey earlier this year, and reached a somewhat different conclusion: 'We are fast heading for e-mail burnout!'

It turns out that we fell, once again, for the Myth of Technological Replacement. 'The original promise was that the new technologies would substitute for the old,' says Steve Woolgar, a sociologist of the internet at Oxford University. 'They don't. They just sit alongside the ongoing means of communication, and it's compounded, because one medium

stimulates the other. The more you do non-e-mail things, the more e-mail you send. So everyone talks about being a member of a different kind of community through e-mail, but then you find people deleting 60% to 70% of the e-mails they receive on the basis of the subject line alone, and you think: what kind of community is it that I'm part of here?'

More and more of us are becoming victims of what the psychologist David Lewis has christened 'information fatigue syndrome'. Symptoms include exhaustion, anxiety, failure of memory and shortness of attention in the face of the uncontrollable onrush of facts. 'Having too much information can be as dangerous as having too little,' Lewis says. 'It can lead to a paralysis of analysis, making it far harder to find the right solutions or make the best decisions.'

Mark Whitby isn't quite there yet, but he knows how it might feel. The 36-year-old UK sales manager for Intel, the dominant multinational manufacturer of computer chips, receives around 100 e-mails on a busy day, and copes by never switching off. 'When I started I was getting two or three e-mails a day, but it's exploded in the last five years,' he says. 'Now, I'll manage my e-mail on the train, on the plane, or download before I leave home and read and reply on the move. We're trying very hard to cut down on using e-mail for small talk, for gossip, for things you could walk over to somebody's desk and say.'

If our brains did not evolve for this, neither did our businesses. The average British workplace was designed on the assumption that information is a scarce resource, employing countless functionaries – switchboard operators, messengers, post-room people, middle-managers, typists and filing clerks – to find it and to transport it from where it is to where it needs to be. But now it is too plentiful, and instead we need filterers. Inevitably, Microsoft researchers are already hard at work on their answer to this problem, a 'digital butler' which will sort and prioritise messages, scanning e-mails for important information such as the dates of meetings or the name of your most feared superior.

At first, the unfiltered nature of e-mail promised an unprecedented democratisation in communications. Managing directors might have been protected from face-to-face visitors by secretaries in anterooms, their mail opened and sifted by assistants, but in the early years of widespread e-mail use, an e-mail often reached them directly. But it was not to last: the invincible indicators of status are catching up with technology. Computer illiteracy as a sign of stature may be on the wane, but, says Woolgar, 'More and more senior managers are having their secretaries open their e-mail, so if you send something to the chief executive, much as there's a public performance that this person is available, of course he's not really going to read them.'

In subtler ways, too, our responses to information overload serve to demarcate status and maintain the old ways of working. David Owens, a professor at Vanderbilt University in Tennessee, had almost finished an anthropological study of status indicators in a major US corporation when somebody suggested he study the firm's e-mail records. Senior types, he discovered, specialised in curt, often misspelt responses, ignoring the rules of punctuation and revelling in lower-case letters like they were going out of fashion. The message was: they were too busy to do anything else. Middle management, by contrast, chose to spell correctly, explain at length, and punctuate.

'Not being accessible generates a powerful symbolic sense of value,' says Owens, who receives more than 70 e-mails most days during term time. 'But the rest of us are going to have to work out some quick ways to filter. If I get an e-mail and I see that the sender's address is webtv.com, well, I'm not going to pay much attention. But if it's from Bill Gates, you know, I'm probably going to read it . . . There is a real struggle here. It's great to have an infinite number of connections, but that's also the problem – you have an infinite number of connections.' (There are more mundane obstacles, too. 'I can't give my password to my secretary to check my e-mail, because the IT administrators told me I couldn't,' Owens says.)

But long before our brains or our organisational systems buckle and die with the stress of knowing too much, it may be the physical stuff on which the internet relies – the servers and wires and switches – that gives out first. At least until broadband and fibre-optic technologies are far more widespread, we are facing meltdown. The tale of the schoolteacher who has their class send a chain e-mail around the world, asking recipients to reply to it and to forward it in order to show the power of the medium, is an often-recurring piece of internet folklore. The school's e-mail system – and that of the entire local education authority – usually holds up for a day or two before collapsing under the weight.

For years now, a petition has circulated on e-mail offering recipients the chance to express their outrage at the treatment of women in Afghanistan by the Taliban regime. They add their name to the bottom of the list and send it on to their own friends, in ever multiplying forwarding trees. Every time 50 names are added to one version of the list, the recipient is asked to forward it to an e-mail address at Brandeis University in Massachusetts – sarabande@brandeis.edu. But the e-mails never arrive: for two and a half years the e-mail address has been de-activated to avoid paralysing the entire university's e-mail system. 'Please do not respond in any way,' the university's solemn warning reads. 'The user of that address no longer wants the mail.'

Knowing the dangers

Information from the Internet Watch Foundation (IWF)

What are the dangers?

If you don't know what you should be worried about, this article explains the risks.

When your child logs on to the Internet, he or she is linking up to a vast network of computers which extends all over the globe. Children can enjoy worldwide access to material that is educational and entertaining, and to services that enable them to communicate with and learn about people from other countries and cultures.

The Internet neither belongs to nor is controlled by any one person, organisation or government. This lack of central control is sometimes portrayed as anarchic and dangerous. But on the whole it offers huge benefits. Unlike any other medium, the Internet gives everyone the chance to be a publisher – to create and publish material for people to see across the world. This is an unprecedented creative opportunity for adults and children alike.

The flip side of this is that other people may use the Internet in ways which you may find offensive, or which are actually illegal. In using the Net you need to know about the possible hazards, particularly for children, and how you can avoid them.

But most material on the Internet is legal, so people have a right both to publish it and to access it. At the same time you have the right to control your own use of the Net by choosing what you personally don't want to see, or don't want your children to see. Software tools are available to help you select the material you access, particularly on the World Wide Web. In addition, Internet Service Providers (ISPs) do vary in the range of services they offer to their subscribers, particularly

in areas such as newsgroups and chat, so you can choose your ISP accordingly.

The Net is not a legal vacuum – the law applies online exactly as it does offline, and people who break the law are subject to investigation and prosecution. In a growing number of countries specialist hotlines have been set up to tackle the problem of illegal content.

The main dangers people are concerned about can be grouped into:

Contact – the greatest danger in the virtual world is letting online contact lead to a meeting in the real world with someone who is not what they pretended to be and who poses a real physical threat. Young people must re-learn the old stranger = danger messages in a new context and use the anonymity of the Net to hide their real location.

Content – legal or illegal, there are some sorts of content that might harm younger users, or that offend the values and standards that you want to apply to your children's development. You can agree the ground rules about where your children go and how they behave, and perhaps choose some software tools to help apply your rules.

Commerce – with the growth of e-commerce there are increasing concerns that in an unregulated global market place young people (and adults too!) may be exploited by dubious marketing practices or simply cheated out of their money.

• Here in the UK, the main hotline is on the Internet Watch Foundation's website. If you want to report something you think is illegal, visit their website at www.internetwatch.org.uk

Internet safety

Children and the internet

The Internet is now part of everyday life for children – whether it's e-mailing their friends or surfing for clues to homework problems. Recent figures suggest that more than 65% of children aged between seven and sixteen are now accessing the Net – and the numbers are going up all the time.

But it is important to recognise the potential risks. Children can find unsuitable sites and there are potential dangers of children 'meeting' strangers over the Net, especially in chat rooms where people are able to pretend they are somebody else.

Parents need to make themselves aware of the realities of the Internet, because even if they do not have access in their own home, their children are certain to be using the Net at school, at friends' houses, in the library or Internet cafés. Parents need to know that the Internet is a vast and largely unrestricted world that has many areas suitable only for adults. There could be over 100,000 chat rooms available to users in the UK at any one time, for example, and while some Internet Service Providers may claim that they offer protection to children, those claims have been challenged.

Teachers should be able to offer advice – it's worth asking what your child's school is teaching about Internet safety – but as always, other parents are probably the best sources of advice.

There are plenty of software 'guardians', sometimes referred to as filtering programmes, that can be downloaded to prevent children coming across unsuitable sites; they screen out unsuitable material and some of them can also both monitor and limit what children can do. Setting up such a filtering tool on your machine is probably preferable to banning surfing or certain sites – that may only make them more attractive. Especially with younger children, it is often a good idea to have the computer in a family room, so that parents can keep an eye on what sites their children are visiting. If they want to go into a chat room, go with them, or at least monitor the conversation.

If a child is old enough to use a computer, then he or she is old enough to understand the possible dangers – especially from chat rooms, or other interactive sites where they may end up 'talking' to others. There have been many stories in the press about children being contacted by people who may pretend to be children themselves.

Parents need to make themselves aware of the realities of the Internet, because even if they do not have access in their own home, their children are certain to be using the Net

Parents are the best judges of their children, and all children develop differently. There are no hard and fast rules about the kind of material children can deal with at what age. But in the same way as any parent warns their children, from a very young age, about talking to, or going off with strangers, any child using the Internet needs to be made fully aware that they must be very careful about where they go on the Net and whom they contact. It's vital that they do not give their name or address, or details of their school, or indeed any personal information like that to someone they only 'know' online. You should also encourage them to tell you immediately if they come across anything or anyone who worries or upsets them, and not feel guilty about it. Try to let them know that you will understand that it's not their fault if other people send them bad 'stuff' or suggest inappropriate things otherwise there is a risk that they'll bottle it all up for fear that you might stop them using the Internet altogether.

It's far better to educate children about potential dangers and how to avoid them than to lay down draconian rules, or make them feel they are being spied on all the time.

Useful contacts

The charity NCH has published a good guide on Internet safety for parents. Visit the guide online at http://www.nch.org.uk/internet/

NCH's checklist for the netsmart

Read – remember – respect

1. Don't Dish it Out!
Never give out personal information, stuff like your address, mobile or home telephone numbers or school names – unless you get the OK from your Parent/Carer.

2. Don't Misplace Your Face!
Keep your ID pictures and banking details to yourself – unless your Parent/Carer says it's OK. Double-check each time.

3. Backup!
Never arrange to meet with your key-pals in the real world without your Parent/Carer and stay in a public place: your web-pal might not really be who they said they were.

4. Attack of the Attached!
Viruses can spread through e-mail attachments when you open them, destroying your computer. Yikes! Make sure the e-mails are from people you know and trust . . . even this isn't foolproof so if you're unsure – bin it.

5. Keep the Peace!
Always check that your Parent/Carer is happy for you to enter a chat room.

6. Word Up!
Your password is PRIVATE – keep it to yourself

7. Web-Wasters!
Don't put up with bullying! If you receive Bad-taste & Bad-attitude messages on the Internet (or e-mail) report it to your Parent/Carer as well as your ISP (Internet Service Provider) – they can help to block unwanted e-mails or get you a new e-mail address.

8. Dodgy Signal?
If you are bullied through your mobile phone (and receive calls or messages that scare or upset you), keep a record of when you got them and tell the police. Talk to your mobile service provider (the company you pay for calls) about maybe getting you a new number. Always be careful who you give your number out to.

9. Wise-Up!
You can't always be sure it's only children in your chat room if you can't see them. Chat safely: it could be an adult winding you up.

10. 'Pop Ya Collar!'
Leave a chat room the moment anything worries you. Let your Parent/Carer know what's up.

11. Be True 2 Yourself!
Don't pretend to be someone or something you're not. Playing games is one thing, but seriously misleading someone about who you are is bad news and can get you into hot water.

12. Reality Check!
Keep clear of Over18 sites. The warnings are for your protection and adult sites can sometimes add serious damage to your phone bill.

Have fun and stay netsmart!

• The above information is from NCH's web site which can be found at www.nch.org.uk

Wise up to the net

Keeping your child safe on the Internet. Tips for parents and carers

What are the issues?

The Internet revolution has transformed all of our lives, rewriting the rules about how we communicate with each other, how businesses operate and how we conduct our daily lives. The Internet has given young people, in particular, a virtual playground, an international school and a place to meet and make friends. More than any other group, young people have embraced the new technologies and made them their own – at school, at home and with friends.

But, sadly, the Internet – like so many other technological advances – is not immune from criminal abuse, and can bring its own dangers, not least to our children. Behaviour on the Internet is subject to the same rule of law as the real world. And, just as in the real world, we need to take precautions on the Internet to protect ourselves from harm.

The Government's Task Force on Child Protection on the Internet is a positive partnership of experts from the Internet and computing industry, child protection organisations and the police, which is working to make the UK the best and safest place in the world for children to use and enjoy the Internet.

We all need to ensure that taking sensible precautions to protect ourselves and our children online should become as commonplace as it is to lock our doors or not talk to strangers in the offline world.

As parents and carers, you have a crucial role to play. Whatever your Internet experience or expertise, this article is not intended to alarm you, but to alert you to the potential dangers that children may face online and help you to help them surf in safety.

Why do I need to worry about safety? I thought the Internet was a wonderful thing?
The Internet is a wonderful thing. It's fun, it's a great way for children to keep in touch with friends, and its educational possibilities are almost unlimited. But like any technology, it can be abused.

How could that happen?
Places called Internet chat rooms allow people to 'meet'. These are areas of the Internet where people have 'conversations' (usually typed, rather than spoken) about common interests such as music, football or television programmes. Everything that is typed can be seen, more or less instantaneously, by everyone else using the chat room on their own computer.

That seems harmless enough. What's the problem?
Chat rooms can be completely harmless and they can be a lot of fun. They are certainly very popular, and there are hundreds of thousands of them. But you can't tell who anyone is in a chat room. Because of this, and because chat rooms are particularly popular with children and teenagers, there is a small risk that they can be used by paedophiles or sex abusers looking for victims. Adults who want to exploit children might pose as teenagers themselves, try to strike up a friendship and eventually, try to meet a child or teenager. This is why the first rule of chat rooms is NEVER to reveal any personal details – that includes full name, address, telephone number, e-mail address and mobile number.

But how can I get involved in all this? I feel completely overwhelmed by the Internet and my children know much more about it than I do.
You can do a great deal. You can certainly be a good parent or carer and teach safety issues even if you don't know exactly how the technology works. Take an interest in what your child does online and if you don't know how to use the Internet, ask your child to show you. Let them know they can come to you if something upsetting does happen. You can also get across the most important safety message about chat rooms. That is, everyone your child meets in chat rooms is a stranger – and remains a stranger even though your child may consider them to be a friend. Surfing the Internet from the comfort and safety of home can give a false sense of security.

Isn't all this a bit far-fetched?
The risk is small but it is real and the consequences can be very serious. In October 2000, a 33-year-old man was imprisoned for having sex with a 13-year-old girl he had met in a chat room. He had built up a relationship with her over several months – a process known as grooming – before they finally met. There have been many other cases, and those who try to contact children online may be abusing children offline as well.

I still don't understand how it can happen. If chat rooms are public places, available to all, everyone else would see what was going on.
Once you are in a chat room, you can be invited to have a one-to-one conversation with someone. This is like stepping out of a party full of people into a private room and having a separate conversation with a stranger. No one else can read what is being written. It could be extremely dangerous. Getting a child on their own, and building up a relationship, is exactly what a paedophile wants. So discourage your child from having one-to-one conversations. They

should stay in the public area of the chat room, which is open to all, and where they should be much safer.

So that's why they shouldn't give out any personal details?
Exactly. Tell your child they should never reveal information about their name, address, password or school. Sometimes children think it's OK to give out a mobile phone number or their e-mail address. It isn't. It gives a stranger direct access to your child, and you have no way of knowing who is contacting them. Even something that seems harmless, like who they're playing hockey or football against next week, or where their favourite pizza restaurant is, could be a clue to their identity. Just as they wouldn't give personal details to a stranger in the street, so it should be with people they meet in chat rooms.

What if they become so friendly with someone they chat with online that they want to meet them in person?
Then always go along too, and arrange to meet in a public place where there are lots of people around. Children and teenagers should NEVER arrange to meet anyone they have encountered online without a responsible adult being present.

What else can I do to keep my child safe?
Advise them against opening links to other sites they might be sent in a chat room (they may be pornographic). Similarly, they shouldn't open e-mails from anyone they don't know (again, they may contain pornographic or upsetting images, or viruses which could harm your computer). This could also be a way for the sender to discover personal details about yourself.

If they are being pestered by someone they don't like in a chat room, there will probably be a facility to block messages from that person.

What if something unpleasant happens while they are actually in a chat room?
They can leave the chat room. Alternatively, chat rooms usually offer the option to block messages from other users. So if someone repeatedly types things your child doesn't like, they can block them from their computer screen. Some chat rooms also let you report abusive behaviour by clicking on an appropriate link.

How do you tell if a chat room is suitable?
Chat rooms are given names, such as 'Teenage Romance'. Teenagers should be encouraged to use only those rooms that are appropriate to their age. This is because so-called 'adult' chat rooms are sexually explicit, where people indulge in what is known as 'cybersex'. This can be anything from talking dirty to the exchange of explicit or pornographic material, including the use of images on webcams. Although it's not necessarily illegal, it is definitely an area for adults only.

How can I encourage my child to use appropriate chat rooms?
At the top of your Internet browser, which is the technology that lets you navigate the Internet, there will probably be the word 'Favourites'. This lets you add to a folder web addresses that you often use, or don't want to forget. You could save the address of child-friendly chat rooms here, and agree with your child that he or she will use only these.

Sometimes children think it's OK to give out a mobile phone number or their e-mail address. It isn't

Wouldn't it be better to avoid using the Internet altogether?
Not at all. Despite the potential for problems, it really is an amazing tool for learning and information. Knowing how to use it effectively is increasingly important and necessary. What's more, chatting online, or Instant Messaging can be a way to cut your phone bills, if you have free or low-cost Internet access in the evenings or at weekends. If you overreact, your child may clam up or simply use the Internet elsewhere. Just drive home the safety messages so your child can feel confident about protecting him or herself.

If I see something illegal or suspicious, is there anyone I can report it to?
Yes. If you think your child is being contacted by a paedophile, call the police. If you come across material you think is illegal, for example child pornography, contact the Internet Watch Foundation. This is an industry-funded body which seeks to have illegal material removed from the Internet and refers it on to the police. The telephone hotline is 08456 008844 or report it online at www.iwf.org.uk/hotline/report.htm Alternatively, contact the police Child Pornography Information Line on freephone: 0808 100 0040.

Why don't ISPs (Internet Service Providers) do something about chat rooms?
ISPs provide the connection to the Internet, just as a telephone company provides access to the phone lines. They can't control what other companies put on the Internet, any more than a telephone company can control what people say on the phone, but some ISPs run their own, sometimes more regulated, chat rooms and they and other chat service providers should be happy to explain any safety features which their services have. Don't be afraid to ask.

What about schools? Don't they teach children about Internet safety?
Throughout the UK, pupils are taught how to use e-mail to communicate and how to evaluate different sources of information

(including websites). They are also encouraged to appreciate the need for responsible use of these technologies in order to protect information, individuals and society. Schools build key safety messages into these topics, with a view to making sure that pupils understand the safe behaviours to adopt when online.

It is important for you to let your child know that you are aware of these safety messages and to reinforce them at home.

What about chat rooms that are supervised?
These are called 'moderated' chat rooms and may use a real person or special technology to block personal details, and keep the conversation appropriate.

They sound like a good idea.
They are a step in the right direction but the technology isn't foolproof and if chat rooms use human supervisors, ask the provider of the chat rooms how they have been recruited and trained. Even if the chat room is moderated, the same safety rules apply.

It all makes me feel powerless and worried . . .
It shouldn't. The important thing to remember is that you can help your child be in control. Children like to feel they have tools to cope. If you explain the safety messages, you are giving them something useful and valuable. Though the Internet is a new medium, the safety messages will be familiar. Just as you have taught them about talking to strangers in the real world, so you can teach them here.

Are chat rooms the only places where they can make contact with strangers?
No. They can also meet people online in places called newsgroups, communities, groups or clubs, and by using Instant Messaging. But the good news is that the same rules apply: NO personal details; NO meeting up with anyone in person unless they are accompanied by an adult.

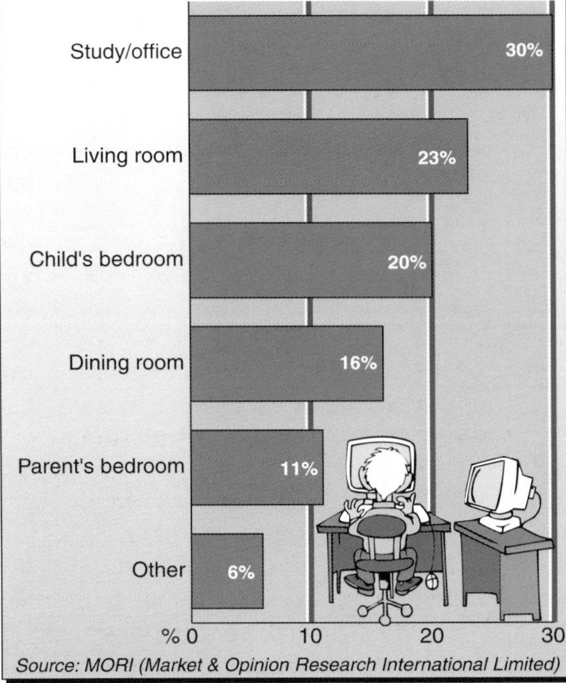

Are there danger signs I can look for?
A child spending an excessive amount of time online could be a worrying sign. Only you as a parent or carer can decide what is excessive, but if it dominates their social activities, you may decide it is too much. Talk of a 'boyfriend' or 'girlfriend' they have met online may also be a cause for concern.

What about pornography?
Pornography is very easy to find on the Internet, even by accident. Encourage your child to tell you if they find something online that disturbs them. Reassure them that it is not their fault and that they will not be punished as a result.

But teenagers are naturally curious about sex. They might go looking for this.
They might. Every family is different, and only you know what your teenager is mature enough to see.

I don't want them to see any of it.
You can buy filtering programmes which aim to exclude access to undesirable material. The Internet Watch Foundation has more information. Ring 08456 008844 or see its website: www.iwf.org.uk/safe/tool.htm. This links (i.e. connects) to a review and a rating of several products by *Which?* magazine. However, no filtering product is 100% foolproof. Savvy teenagers can by-pass some filters or use the Internet, unfiltered, in a friend's house or a cybercafé. Filters have their place, but they are no substitute for good parental guidance.

What about moving the computer?
If possible, it's an excellent idea to keep it in a family room.

Where else can I get help?
Your Internet Service Provider should also be able to provide safety information and tell you what services it provides for children and teenagers.

Who can help me?
- Internet Watch Foundation: Industry-funded body with hot-line to report illegal material. Tel: 08456 008844, www.iwf.org.uk
- Police Child Pornography Information Line: Freephone police number to report incidents of child pornography. Tel: 0808 100 0040. This can also be accessed via www.met.police.uk
- For Kids By Kids Online: Independent body based at the Cyberspace research unit at the University of Central Lancashire, www.fkbko.net
- Childnet International: Not-for-profit organisation aiming to make the Internet a great place for children. www.childnet-int.org. Its related site, www.chatdanger.com, has specific safety information on chat rooms.
- More information about the schools' Internet Proficiency Certificate for Key Stage 2 pupils (7-11-year-olds) is available online at www.becta.org.uk

• Our thanks go to ChildLine, Childnet International, the Internet Service Providers' Association and the Internet Watch Foundation and the NSPCC for their help in producing this article.

Internet filtering software

Performance tested

Filtering and blocking access to 'undesirable' content or contact on the Internet is obviously something many parents wish to be able to do. The objective of these evaluations is to help parents with an interest in managing their children's Internet usage to decide which, if any, of the currently available Internet safety software will best suit their needs.

This software (often called filtering software) offers parents a range of methods of controlling, monitoring, filtering or blocking access to the Internet.

Our evaluations show that, whilst most products offer some level of protection, none can be relied upon by parents as a complete solution. Strategies for protecting children might involve the use of a filtering software product, but should also include agreements with children about online times, unsuitable content and unsafe online contacts. Children will also need strategies to cope when the filtering software does not work or is not present.

Overall findings

1. There is no agreed standard way to solve the problems of monitoring and protection. There is a huge variation in the rationale behind these different products and in the methods they offer.
2. No single product was found to combine all available filtering features.
3. There is no agreed quality standard of technical performance or user support.
4. When tested, most programs did not work as effectively as advertised.
5. Different kinds of families need different kinds of monitoring and protection and so must search for products which suit their needs.
6. Products will need to be updated regularly to cope with changing patterns of Internet use by families, changing content of websites and new ways to access online resources. None of them provides a permanent solution for parents.

The evaluation process

Products were evaluated by PIN using its national network of families over a six-month period. Parents were asked to use the products and complete a set of detailed questionnaires based on the evaluation criteria below.

Products were also subjected to a rigorous performance test designed by PIN to establish their effectiveness in preventing access to unacceptable content such as nudity, sex, violence, extremism, racism, drug use, bomb-making, in websites, e-mail, chat, newsgroups and search engines.

Criteria:

Ease of installation – how easy is it to install without interfering with existing software?

Ease of configuration / updating – how easy is it to configure the program settings and to incorporate updates?

Clarity / transparency – does the 'administrator', i.e. the parent, know what has been filtered or blocked?

Flexibility – to what extent can the software be adapted for different family members?

Effectiveness – how well does the program filter or block the content it claims to filter or block?

Security – how well protected is the administration area from unauthorised tampering?

Support – how much user support is provided and how supportive is it?

This evaluation project is a huge undertaking and more information is still being processed. There are some areas which need further testing and evaluating. We are committed to providing the best possible advice to parents as purchasers, as well as to publishers in terms of product development to ensure that the needs of families are met in this key and rapidly changing area.

This software offers parents a range of methods of controlling, monitoring, filtering or blocking access to the Internet

• The above information is from the Parents' Information Network's web site which can be found at www.pin.org.uk

Kids look here!

Safety guide

NBD . . . No Big Deal

Let's face it – when it comes to the Net, we know much more than our parents. It's No Big Deal being able to surf the Net, be in a chat room and txt your mates on your mobile all at the same time!

Who R U?

You might think you've met some real cool friends in chat rooms but how can you be sure who you're talking too? The person may be nice and trustworthy, but then again they may not be. I tell my mates not to put too much info in my chat profile and never arrange to meet up with anyone you've been chatting to – chatroom mates are best left in cyberspace. You might think you're clued up about using the Net safely, but make sure you look out for your mates and remind them to be very careful and stick to the SMART rules in this article.

Stick to the positive

The great thing about the Internet is that there are some fab sites out there – many of which have been created by young people! If you find a really cool site which helps you with your homework or hobby tell a friend. Nasty stuff like hate sites, sectarian or pornography are right out! Get a life! – Leave those sicko sites where they belong – in the trash can! Be particularly careful about sites which ask you for lots of personal details! You never know where that information goes!

Mobile up!

Everyone's texting mad – but it's not only your mates who are buzzin your fone! Don't give your mobile phone no out to anyone you don't know and if you start getting annoying or offensive texts talk to your parent or guardian straight away. Keep your mobile out of sight when you're on the street and make sure you have a security code or PIN to lock your phone.

Keeping smart online!

There are some really cool things on the Internet but a lot of bad stuff too. This means we have to be SMART when we are online. See if you can remember these 5 Safety Tips, and then prove to your friends and parents that you are a SMART surfer.

SECRET – Always keep your name, address, mobile phone number and password private – it's like giving out the keys to your home!

MEETING someone you have contacted in cyberspace can be dangerous. Only do so with your parent's/carer's permission, and then when they can be present.

You might think you've met some real cool friends in chat rooms but how can you be sure who you're talking too? Chatroom mates are best left in cyberspace

ACCEPTING e-mails or opening files from people you don't really know or trust can get you into trouble – they may contain viruses or nasty messages.

REMEMBER someone online may be lying and not be who they say they are. Stick to the public areas in chat rooms and if you feel uncomfortable simply get out of there!

TELL your parent or carer if someone or something makes you feel uncomfortable or worried.

- Childnet has created a special website all about chat rooms called www.Chatdanger.com. This website tells what happened to one family when their daughter met up with a stranger she met in a chat room. The site also includes examples of where chat can be used positively and what to look for in a good, safe chat room.
- The above information is from Childnet International's web site which can be found at www.childnet-int.org For examples of positive ways children are using chat also see www.childnetawards.org

Overview of the problem

Information from the Internet Hotline Providers in Europe Association (INHOPE)

The Internet is a wonderful medium for education, research and entertainment and, indeed, for the conduct of business.

This can sometimes be forgotten when we are looking at the downside issues on the Internet.

Without resorting to extreme, inappropriate and inevitably unworkable measures of censorship and restriction, the Internet will always include users who represent the darker elements of our society.

INHOPE believes, therefore, that it is not a question of guaranteeing the removal of all illegal and harmful material on the Internet or of guaranteeing the impossibility of access to such material.

Rather it is a question of working towards a safe environment for Internet users which will protect our children and respect the privacy and dignity of our citizens.

To achieve an understanding of what we can do to address the downside of the Internet, we must first confront and understand the nature of the Internet. The key areas of concern are:

- Child Pornography
- Commercial Sites
- Morphed/Edited Images
- Chat Rooms/ Abduction
- Paedophile Rings
- Racism
- Adult Pornography
- Grooming

Child pornography

What it is

Child pornography has different legal definitions in different countries. The minimum defines child pornography as a picture that shows a person who is a child and who is engaged in or is depicted as being engaged in explicit sexual activity.

The European Commission Green Paper draws a distinction between material which is illegal and unsuitable for viewing irrespective of the viewer's age, and that which is undesirable for children:

Many people find it difficult to imagine pornographic images of children, and therefore do not understand what is meant by 'child pornography'.

Also, among the issues which cause disagreement are the age of consent to sexual relations, whether simple possession of child pornography should be a crime, whether an actual child had to be involved and whether morphed images constitute pornography.

How the Internet is used

It must be remembered that apart from the obvious damage to the children who are actually used in the making of child pornography, there are dangers in children getting access to material which may be pornographic, hate-based, violent or in some way threatening to the well-being of the child.

Images in a range of file formats can be stored on and transmitted between computers

A major concern is, therefore, the ease with which pornographic material can be accessed, stored and disseminated on the Internet.

The term 'child pornography' is, in itself, not fully descriptive of what is involved.

Assuming that the images are

not computer-created, images of adults engaged in sexual activities with a child are nothing less than images of an assault on that child.

Images in a range of file formats can be stored on and transmitted between computers using the Internet. Such images can be created through scanning of photographs/ slides or digital-capture on video at relatively low cost, converted into a standard image file format and displayed or transmitted to other like-minded individuals.

Commercial websites

What it is

Child pornography websites are very rarely commercial in nature and are motivated by personal gratification and the need to seek out like-minded persons. This can be seen in the case of the wonderland club where the membership 'fee' was to exchange thousands of pictures.

However, there is a worrying development in recent times with the growth in commercial websites requiring payment of membership fees usually by credit card or other financial methods.

These sites tend to be located in legal jurisdictions which are not sufficiently up to date with regulations governing such activity.

In addition these sites are often located in lower economic areas where the value of human life is not regularly respected.

How the Internet is used

Websites can be hosted in any location in the world (indeed in multiple simultaneous locations making up the jigsaw pieces of a complete website). It is possible to have password or controlled access to websites.

These access passwords are sold on the Internet with subscriptions based on a monthly or yearly fee. They can be purchased remotely using credit cards (sometimes stolen) for remote access.

Once the password is available, the website can be accessed in the normal way from any internet connected machine in the world

Fake/morphed/edited images

What it is

Computer images can be created, edited and changed.

This can be small subtle changes or major changes. Images can be created from blank such as cartoon-generated images.

It is also possible to change a picture of a naked man in a sexually explicit pose by replacing the head of the man with that of a boy.

This gives the impression of a boy being involved in explicit sexual activity.

How the Internet is used

Images in a range of file formats can be stored on and transmitted between computers using the Internet. Such images can be created through scanning of photographs/slides or digital-capture on video at relatively low cost, converted into a standard image file format and displayed or transmitted to other like-minded individuals.

These images are then changed (morphed) or edited to create a certain required look.

Chat rooms/abduction

What it is

This 'real-time' service offers paedophiles the opportunity to discuss their fantasies with others and to share experience in an anonymous context on the Internet.

The main service, Internet Relay Chat (IRC) is also used to arrange face-to-face meetings. While most IRC sessions are public, private sessions are possible and there have been a number of reported instances of known paedophiles using such interactive sessions with children to gain their confidence before attempting to arrange meetings.

How the Internet is used

A hybrid arrangement somewhere between e-mail and IRC is emerging in the form of ICQ services which allow private communication to take place in real-time mode between parties subscribing to a particular directory.

These services are known to be used for paedophile purposes and are, because they leave no electronic footprints, very difficult to detect.

Paedophile rings

What it is

A paedophile ring is a group of persons working together across the Internet in different countries and jurisdictions to collect and distribute child pornography for their own gratification.

This can also involve sharing expertise and experiences on avoiding detection and planing criminal activities against children.

How the Internet is used

There is a strong perception that the Internet has become a major factor in the development of paedophile rings world-wide and many recent convictions in the United States and in the United Kingdom have shown that the medium is being widely used by members of such rings, both to share experience and to traffic in child pornographic images.

The dissemination of child pornography is causing major concern to the international agencies engaged in the protection of minors. Irrespective of the routes used for disseminating child pornography the problem continues to be serious in Western Europe where major child pornography rings have been uncovered in Denmark, Germany, Italy, the Netherlands, Sweden and the United Kingdom.

As these networks increasingly use advanced telecommunications technologies, making use of encryption and code names, they have become more and more difficult to uncover.

Late last year, the head of the Internet Crime Forum said that 20 per cent of British kids using chat rooms had been approached online by paedophiles.

> ***The head of the Internet Crime Forum said that 20 per cent of British kids using chat rooms had been approached online by paedophiles***

Racism

What it is

Investigating Internet content which might be judged racist related or hate-speech related is extremely complex and is a new area of concern. However offensive some of the material might be, it does not tend to fall foul of the criminal law.

On Tuesday 27 June 2000, German Justice Minister Herta Daeubler-Gmelin called for global rules against hate-speech on the Internet and urged stronger self-regulation by web companies to beat racism and xenophobia.

How the Internet is used

In Germany the number of extreme right-wing homepages has jumped to 330 in 2000, about 10 times more than four years ago, the country's internal security watchdog says.

Germany is especially wary of the abuse of the Internet given the country's Nazi history, but everybody needs to be on their guard.

The German Justice Minister said Internet companies had an obligation to make sure they did not sell extremist books or music, while service providers could block websites promoting hatred.

The German Justice Minister welcomed the fact that Internet retailer Amazon.com Inc had agreed to stop selling Adolf Hitler's outlawed *Mein Kampf* into Germany, but said the company should not sell the book anywhere in the world.

Adult pornography

What it is

Pornography has different legal

definitions and different levels of acceptability in different countries.

The minimum defines pornography as a picture (video, dvd, audio, text) that shows a person who is an adult and who is engaged in or is depicted as being engaged in explicit sexual activity.

The European Commission Green Paper draws a distinction between material which is illegal and unsuitable for viewing irrespective of the viewer's age, and that which is undesirable for children:

It is important to distinguish two types of problem relating to material:

Firstly, access to certain types of material may be banned for everyone, regardless of the age of the potential audience or the medium used.

Here it is possible, irrespective of the differences in national legislation, to identify a category of material that violates human dignity, primarily consisting of child pornography, extreme gratuitous violence and incitement to racial or other hatred, discrimination and violence.

Secondly, access to certain material that might affect the physical and mental development of minors is allowed only for adults.

How the Internet is used

Images in a range of file formats can be stored on and transmitted between computers using the Internet.

Such images can be created through scanning of photographs/slides or digital-capture on video at relatively low cost, converted into a standard image file format and displayed or transmitted to other like-minded individuals.

Grooming

What it is

A preoccupation with marginal, unpleasant material can often give rise to false and distorted beliefs about the world. Pornography, for instance, relies heavily on equating sexual gratification with physical and emotional abuse, and consumers of it frequently believe that the victims enjoy their torment or are unharmed by it.

Propaganda is believed if it is accompanied by plausible arguments or satisfies individual emotional needs. Correspondence with others who share beliefs greatly adds to their credibility. The intellectual limitations, emotional liability, immaturity and inexperience of children and young people make them particularly vulnerable to all this.

Engrossment in any medium will affect a person's values and attitudes. Children and teenagers can be particularly impressionable as they are still in the process of developing theirs.

Paedophiles use the rooms to groom children for abuse by pretending to be teenagers themselves

Intense use of undesirable material can create negative mindsets or exaggerate already established personality traits. Pessimism, personal antagonism, cynicism, unpleasant precocity, disaffection, diffidence, sexism, racism, intolerance and anti-social posturing are some of the possible outcomes.

A study entitled Chat Wise, Street Wise released in March 2001 by the Internet Crime Forum in the UK, found five million young people are on line in the UK and a quarter of all children use chat rooms, a special kind of website where they can have typed conversations with strangers.

Paedophiles use the rooms to groom children for abuse by pretending to be teenagers themselves.

Girls aged 13 to 17 are most at risk, such as in the case of 30-year-old Patrick Green from Iver Heath, Buckinghamshire, who lured a 13-year-old girl to his home for sex. He was jailed for five years in October.

How the Internet is used

Chat rooms are hosted on the Internet and can be accessed by anyone from anywhere in the world. Chat rooms are often considered safe by children because of the public nature of the conversations AND the incorrectly perceived anonymity.

Paedophiles initiate conversations with likely victims to develop a rapport with the victim and to elicit as much information about location, interests, hobbies and sexual experiences.

This relationship is then developed into sexual conversations and sometimes to examples of pornography – both adult and child – to instil a sense of acceptability and normalcy. This is then used to undermine the reluctance of children to participate in a sexual encounter. It is also used to prevent the victim from seeking protection from their parents and teachers.

• The above information is from Internet Hotline Providers in Europe Association's (INHOPE) web site which can be found at www.inhope.org

General Internet safety tips

This article is designed to give the novice Internet, e-mail, ICQ and IRC (chat) user some rules to help them stay safe while enjoying the Internet

General rules to help you and your children stay safe:

1. Never give out personal information on the Internet!
2. Never give out personal information on the Internet!
3. Remember rules one and two!

Okay, laugh if you want – but that is the single most important rule for staying safe while using the Internet.

Now, then, what is considered personal information?

Your real name, address and phone number, where you work or attend school, what sports teams you play in or band you are in.

How old you are or whether you are a man, woman, boy or girl.

The names of your friends or relatives and any of the above information about them.

The reason for this will become apparent as we progress through the various applications.

Now let's take a quick look at the individual Internet applications you might want to use. We will start with IRC. IRC = Internet Relay Chat!

In a chat room you may have someone ask you for asl. This is a shortcut method of asking you your age, sex and location. Never give out that information. Why?

In any chat room – including our classrooms – you never know who is really in that room with you. You cannot see the person on the other side of the keyboard that is chatting with you.

You don't know who is lurking in the background, taking notes, looking for their next victim.

Lurking in chat rooms is a favourite way for paedophiles to gather information on children. They pick up the little titbits until they know enough to make online friends with the child – and then . . . anything might happen!

Maybe the nickname you know and are talking to went to bed and their brother, uncle, father or a friend is using the computer and talking with that person's nick.

If you follow rules 1, 2 and 3 above – it will not matter whom you chat with!

Let us look at another way people can find out personal information about you online.

When you set up your IRC client or ICQ account you have a profile to fill in which asks all kinds of questions like age/sex/location/real name/address/hobbies/phone number and so on.

Remember the rules – don't put down your real information. Most of these boxes can be left blank.

If they cannot be left blank then don't give 'real' information. Including your e-mail address and nickname.

What is so bad about your e-mail address? Lots of people use their real name in their e-mail address. For example johndoe@ohio.com.

It is easy to trace e-mail and from the above example we know then the person is John Doe, lives in Ohio and by doing a simple search would find that this John Doe lives in the Akron, Ohio area.

Now it is a simple matter to go to an online phone directory and look for John Doe in the Akron, Ohio area. If the last name is not a common one John will be especially easy to find!

The online phonebook will even show you a map of where John lives and step-by-step driving directions to his house. And the person searching has already found out John has two lovely little 6-year-old twin daughters during his online chats with John!

How about that catchy nickname you might use like ohiogal or sexyteen. They give personally identifying information out to everyone who cares to look.

Now let's take a look at e-mail. E-mail is a lot safer than ICQ or IRC. You know (don't you?) whom you are sending e-mail to.

If you really do know whom you're sending e-mail to, like your college roommate in Wyoming, then talk away. Exchange all the information you want to share. But . . .

If that e-mail you are replying to came from someone you don't know, or maybe is a survey from some company or organisation – follow rules one, two and three!

Never give out personally identifying information.

Now lets take a look at the WWW – worldwide web. Lots of companies ask you to fill out a questionnaire or take a survey when you visit their website. If you feel inclined to do so – fill it out – BUT – follow the rules – never divulge personal information.

Now an exception to the rules – yes that's right, there is an exception to the rule. You want to purchase something from an online store with your credit card. You must give them not only your personal information but also your credit card number.

Is that safe – yes – maybe – no?

Before you send any personal

information including your credit card number, look near bottom right corner of the browser (Internet Explorer or Netscape) window. You should see a small yellow padlock. This padlock should be closed – in the locked position.

If the padlock is closed and locked, then look at the address in the address bar of the browser. This address should start with 'https://' not http://. Notice that little 's' on the end. It means that this is a secure website.

A secure website encrypts the information they send you and the information you are sending them to make it nearly impossible for anyone to see and decipher that information while it is being transferred.

If you follow this simple rule – never give out personally identifying information online – you will have a safe Internet experience!

• The above information is from CyberAngels' web site: www.cyberangels.org

The Net of corruption

Nine out of ten children 'have seen porn or violence while surfing'

By Sean Poulter

Ninety per cent of youngsters surfing the Internet have been exposed to pornography and violence, researchers have found.

The results brought demands for stronger safeguards, including greatly increased policing and tougher laws.

Two research projects have revealed how easily youngsters are exposed to the dark side of the world-wide web.

One, by social psychologists at the London School of Economics, found that nine out of ten children ahead eight to 16 had viewed pornography on the Internet.

The other from the EU, found that youngsters in the UK are more at risk than those elsewhere in Europe. Nine out of ten British children aged 11 to 14 are surfing the Internet and a higher proportion are doing so unsupervised than anywhere else in Europe, it claims.

The LSE project found children usually stumbled across pornography while searching for pictures on innocuous subjects such as the Spice Girls or the White House.

Half the children in the study had been approached by strangers in computer chatrooms, sometimes with a view to illicit sex.

Project leader Professor Sonia Livingstone, who watched children surfing at home, said: 'I observed one long flirtation between a 15-year-old girl and a man of 32 that seemed to be the kind of communication that could easily go wrong.'

The EU study claims that children as young as six are being confronted with explicit sexual images from innocent Internet searches.

It also highlights the dangers posed by child pornographers and paedophiles who use the anonymity of the Internet to seek out victims.

Its recommendations for draconian measures to tackle the menace are expected to be adopted by the European Economic and Social Committee (ESC) this week.

The young victims

The LSE study found that children usually accessed sex sites unintentionally while doing homework. Most were shocked and quickly shut them down – although some searched for them deliberately.

- A girl of 11 seeking pictures of Adolf Hitler for a school project found herself led to a website displaying child pornography under a 'gay sex' heading.
- Two brothers aged ten and 12 wanted pictures of the pop band Boyzone but instead were plunged into a gallery of homosexual images.
- A girl of ten who tried to log on to a Spice Girls website was confronted with pictures of topless women.
- A 13-year-old girl doing a project on the White House found herself looking at a porn magazine site.
- A girl of 15 researching the subject of 'money' ended up at a website about prostitution.

These include greatly increased resources for European police forces, a European Standards Body with legal powers to police the Net, and a rating system for websites similar to the one for cinemas.

The study demands tougher EU-wide laws covering incitement to violence and racism that would be used to shut down anti-social websites.

British consumer research expert Ann Davison, who compiled the research, said: 'Three-quarters of British children are still finding harmful material on the Internet. That is pornography, violence and gambling.'

The report proposes the development of an EU Internet Action Plan backed by new legal powers and the establishment of a European Standards Body to police the web.

Calling for European police forces to be strengthened, it adds: 'Specialist police units must be equipped with the most up-to-date and most appropriate hardware and software for their investigations, so that they are not working at a disadvantage.'

Internet service providers are key to policing the material they broadcast over their systems, the study says. It suggests that they should promote chatrooms specifically for children that are moderated by experts in spotting approaches by paedophiles.

The ESC is an official EU forum for representatives of a wide range of interest groups. Its work generally triggers action by the European Commission.

Making your family Internet code

The Internet

- Is it really filled with perverts?
- Will my child come across pornography?
- Is my child in danger of being bullied or approached by paedophiles?

It is important to know that ALL of this is possible . . . but it is NOT inevitable.

It is rather like all the dangers your children are exposed to when they are out of your care and outside the safety of their home. The difference is that, if you have a family computer, these dangers can come into your home.

Both you and your children need to know what the dangers are and how to cope with them. It is not enough simply to say we won't have the Internet in the house (that is a bit like saying you won't let them out of your home!). The reality is that your children will be accessing the Internet whether or not you have it at home.

In order to help you discuss this fully with your children, we have produced a set of points which cover most of the issues you need to consider. In some instances we have included additional details. This information is not meant to alarm you, although it is indeed sobering. It is designed to give you a clear understanding of what could happen (but this is not to say that it will). Only in this way can you be in a position to help your children avoid potential danger.

There are lots of sets of 'Internet Safety Rules' available from all sorts of organisations, but we feel that the most effective way to protect your family is to develop your own code together. That will be something your children are far more likely to remember.

First things first . . .

Keep Internet-connected computers in a communal area of your home with the screen facing outwards. One of the most important ways to protect your child is to ensure that any Internet-connected computer or games machine is not located in their bedroom. Ideally it should be placed somewhere in the house which is commonly used by everyone, where it is quite normal to pass through and notice what is happening.

Get involved yourself – become an Internet user yourself and even take part in a few chat sessions. Then you will have a better understanding of the way the technology works and it will not seem unusual to your children that you are interested in their online activities.

Acceptable use

Be clear about what you consider to be unacceptable information to look for and use on the Internet – e.g. sexism, racism, violence, bad language, pornography, etc.

Be clear about what is acceptable communication – your child may receive abusive messages from a bully at school or from someone met via the Internet. They themselves may even send messages using bad language to other children because their friends do the same.

Downloading unknown files – it is advisable to agree never to download files from strangers or people you do not trust.

Agree, if possible, who can use the Internet, and when – your Internet-connected computer has the potential to be a shared resource which everyone in the family can use. However, sometimes it is quite hard to manage everyone's need to use the computer and the Internet!

Agree how long each person can be online – explain to children about Internet access telephone costs and any concerns you might have about this. Cheaper rates at the weekend and evenings may help.

Visiting friends and family – do you want your family code to apply to visitors? If so, how will you deal with this?

Personal safety

Emphasise what you have already taught them about 'Stranger Danger'. Children are taught to be wary of strangers and avoid contact with them. Most child predators depend on their victims believing that they are trusted friends. Make sure your child understands that, no matter how many friendly chats they may have with someone online, they remain a stranger. The only people who are not strangers are those they know in the real world.

Explain that passwords, address, pin numbers, credit card details, phone and e-mail details are all private and should NEVER be given to

anyone via the Internet, particularly if that person is only known via the Internet.

Because not everyone on the Internet is who they say they are, your children should NEVER arrange to meet anyone before checking with you. The worst outcome of this could be the result of a process of 'grooming' which online predators use to encourage children eventually to meet with them.

Try to ensure your child knows it is safe to tell you about anything that bothers them so that they don't feel it is their 'fault' (they are more likely to tell you if you have already made a Family Code).

What direct action can you take?

If your child has their own e-mail address it is best if it does not give any indication of their age or gender.

You may wish to consider using software which can filter out or block what is allowed to be shown on your computer and what is allowed to be sent from your computer.

Find child-friendly chatrooms with full-time, trained moderators for your children to use. Ask your Internet Service Provider (ISP) to suggest some. If they can't, ask why not. A child-safe chatroom should include visible warnings, child-friendly guidance, a full-time moderator, an easy way to record and report unsuitable approaches, clear rules of use and clear sanctions against anyone in breach of those rules. See smartparent.org.uk for more information about online chat safety.

Contact your Internet Service Provider (ISP) and find out what child-safety measures they offer, if any, and how to use them. If they don't, you may consider changing your ISP to one which caters more directly for families.

Complain to your Internet Service Provider (ISP) if you or your child find any inappropriate content or are subjected to any inappropriate contacts by strangers online.

In the case of possibly illegal material, contact the Internet Watch Foundation via their web site at http://www.internetwatch.org.uk/

The issues covered in this article are obviously very worrying. They are often the issues which are highlighted in the press. However, it is important to keep these things in perspective.

Whilst the Internet introduces new potential dangers it also brings some really fantastic benefits to children and their learning which need to be balanced against the possible risks.

The areas covered in this article are not necessarily going to affect your child directly, but they are real risks which your family needs to be prepared for.

• The above information is from the Parents' Information Network's web site which can be found at www.pin.org.uk

Know the rules of the cyberroad

Ms. Parry's rules for correct Internet behaviour

People do outrageous things when they get behind a keyboard. Things they would ordinarily never do in real space. Somehow, whether it's the fact that they think they're anonymous, or that the Net brings out personality disorders, I don't know. But please don't fall into the trap of saying and doing things online that you know shouldn't be said or done. Never do anything online you wouldn't do in real life, or say anything to anyone you wouldn't say face to face.

Although many of you are seasoned veterans, a few of you are new to both online services and the Internet. I put together some basic rules of netiquette (proper conduct on the net and online services).

Netiquette

It's always good to know the rules before you set forth in a new country, and the Net is the newest country of all. Remember that people from many countries will be sharing ideas, and the rule of the game is RESPECT.

Don't worry about having to learn all this new stuff . . . we were all newbies once, and you'll soon be a seasoned veteran, laughing at all the inside Net jargon.

- Never use all capital letters, it's considered shouting and is hard on the eyes . . .
- Think before you type! Whatever you say always comes back to haunt you.
- Don't refer to anyone online by their real offline names unless they say it's okay.
- Don't copy other people's material and pretend it's yours.
- Don't do anything online you wouldn't do offline.
- 'Flaming' is inciting or provoking an argument and is bad netiquette.
- 'Spamming' is posting something in many places at the same time and wastes other's time. Spamming is a no-no.
- Don't jump into the middle of a chat, learn the rules of the chatroom or discussion board first, and get to know the tone before you join the discussion. It will prevent lots of problems.
- Don't correct anyone in public if you think they made a mistake, send them an instant message or e-mail instead.

Remember that just because you're hiding out in your room behind a big computer monitor, you aren't exempt from correct and thoughtful communication. And also remember that you aren't as anonymous as you think you are . . .

• This information has been provided to us by its author, Parry Aftab, Esq., Executive Director of the WiredSafety family of sites and programmes, Wiredpatrol.org, wiredkids.org and cyberlawenforcement.org It also appears in the *The Parent's Guide to Protecting Your Children in Cyberspace* (McGraw-Hill 2000).

Policing the net

Peter Robbins takes over as chief executive of the Internet Watch Foundation at a time when the self-regulatory group is facing increasing charges of censorship. Sarah Left asks: how will this ex-cop respond?

By Sarah Left

Peter Robbins admits he does not know much about the internet. Aged 50, he retired as chief superintendent at Hackney police earlier this year, and picked up a copy of the *Guardian*'s Society jobs section.

And there it was: an ad from the Internet Watch Foundation, the self-appointed, self-regulatory body that polices child pornography online.

A cop at the head of the IWF, which is at heart an industry body of internet service providers, is not such a strange fit.

The IWF was set up in 1996 with the dual, but mutually beneficial, purposes of limiting ISP liability for illegal goings-online and ridding the net of paedophiles.

ISPs were worried that if they did not police the contents of their servers themselves, then the real cops would step in, leading to litigation, regulation, and, well, nothing good was going to come of it.

Self-regulation seemed the way to go, removing pages and users that engaged in illegal activities from fraud to race hatred.

The biggest target they identified, however, was the abuse of children and proliferation of paedophile rings.

Robbins has no doubts about his priorities.

'I am focusing on child porn and not on the wider issues of fraud or other illegal activities,' he said.

'The written word is not necessarily contained in the Protection of Children Act; only images are prosecutable. And there are no statutory offences for grooming children [whereby paedophiles use chatrooms to gain the trust of potential victims]. The Internet Crime Forum is looking at that.'

The charge most frequently made against the IWF is that they are non-accountable censors. Most of the controversy has surrounded newsgroups, electronic bulletin boards dedicated to specific topics where users can exchange ideas, stories, news and photos.

One IWF board member, Malcolm Hutty, general director of the Campaign against Censorship of the Internet in Britain, resigned in February after the board voted to shut down newsgroups simply on the basis of their name sounding like a forum for illegal content.

'This is censorship of legal debate by an unelected and unaccountable clique acting in secret,' Mr Hutty said at the time.

Mr Robbins says he understands the controversy, but maintains that groups are not shut down without proper investigation into their content.

'Any group that has child porn on there should get shut down and most responsible ISPs understand that.'

At any rate, he argues, out of 90,000 newsgroups in the UK, only 30 have been taken down, 18 are being monitored closely and 20 are considered suspect.

'It's a relatively small number of newsgroups, so I don't know why people are getting so hyped up about it, to be honest,' he says.

While Mr Robbins – on his second day in the job and still rooting around for statistics – spoke, his staff of five handled a steady stream of calls and e-mails to the IWF hotline.

In the last two days, they had fielded 267 reports of illegal activity online, of which 155 had been dealt with following earlier complaints, another 77 were not illegal, and one was outside the group's remit (it was not child pornography).

But 34 seemed illegal to the IWF team, and would therefore be investigated and passed on to the police.

Mr Robbins considers this the essence of the IWF, to sit between the public, the police and the ISPs and to coordinate responses to illegal material.

If those calls and e-mails were being fielded at a police station, he argues, they would simply fall to the bottom of the pile for officers concerned mainly with theft and violent crime.

Much of the material complained about does not sit on UK servers, he adds. When the IWF was set up, its mission was to clean up what it could on behalf of UK ISPs, thus considering illegal material on foreign servers outside its realm of influence.

Now Mr Robbins considers the partnership with foreign counterparts fundamental. The IWF forwards complaints about foreign sites, chatrooms and newsgroups to the relevant local body.

'They run a cyber tip hotline in America and they are very proactive. There are almost daily e-mails passed between us,' he says.

Child pornography is the scourge of the internet, with Unicef reporting that online violations make up 90% of investigations, 80% involving more than one country.

There have been some notable successes. In February last year, seven British men were jailed for taking part in an international paedophile ring called Wonderland that maintained a substantive online library of child pornography.

The next battleground will be mobile phones, says Mr Robbins, as people begin to access data sites through their handsets and encounter illegal material.

He has already been contacted by 'two major mobile phone companies', he says, who are interested in setting up an arrangement similar to that which exists for ISPs.

He also hopes to build better relationships with television companies, as internet access moves into more homes through TV screens.

One of Mr Robbins' plans is to improve relationships with the country's 43 police forces, and given his background he is in a unique position to understand the best way to move from identifying an illegal site, newsgroup or chatroom user to prosecution.

And as for censorship, he rejects criticism that having ISPs shut down sites or newsgroups without a court ruling amounts to a stifling of free speech.

'That's rubbish. That's not our role. Our role is to identify material that is illegal and to have it removed and to provide evidence to police to prosecute. It's as simple as that.'

Police swoop on international net paedophile ring

Fifty suspected paedophiles were arrested in seven countries this morning, in an international police operation to smash an internet child pornography ring focusing on child sex abuse.

Officers from the UK's national hi-tech crime unit had spent a year trying to infiltrate the 'Shadowz Brotherhood', which posted images of child abuse and tips on avoiding detection by police.

Experienced detectives said the images and videos involved, many showing sexual abuse of babies, were the most horrific they had ever seen.

Of the 50 suspects held today, 31 were in Germany.

Some members of the network were involved in sexually abusing children and then posting images of the assaults online for others to download.

Their website provided advice on how to groom children in internet chat rooms for abuse and how to evade detection.

In the action, codenamed Operation Twins, the homes of 50 suspects were raided and computers, videos and CDs seized. The operation was backed by the European police agency, Europol.

The head of the British team that led the investigation, Detective Chief Superintendent Len Hynds, said: 'This group were using highly sophisticated technical means to continue their criminal activities and to avoid detection.

'It was a level of sophistication we have not seen in law enforcement before.'

Members used advanced encryption techniques when uploading and downloading images to and from the site.

The website was allegedly run by hardcore paedophiles, who called themselves 'administrators'.

After initial vetting by an administrator, new members received a 'one star' rating, allowing them to enter certain chat rooms, newsgroups and bulletin boards.

To gain further stars they had to post images of child sex abuse for viewing and downloading by other members. As they climbed the star ladder, they gained deeper access to restricted sites and password-protected rooms containing the most explicit material.

Detectives compared the structure of the site to a castle with heavily-guarded inner sanctums.

The group operated within a cellular structure similar to that of terrorist and international drug smuggling operations, in which only small groups of people within the network were known to each other.

Police estimated that the ring had about 100 members, of which 23 were administrators.

'They provided advice about police tactics and techniques so they could avoid detection. It's clear they had a reasonable awareness of our tactics and we will be exploring every avenue to determine how they accessed that information,' Mr Hynds said.

'The administrators monitored bulletin boards and chat rooms, ensuring people were using proper security measures, and excluded people from the site if they weren't.'

The group is believed to have members in Britain, America, Canada, Belgium, Denmark, Germany, Italy, the Netherlands, Romania, Spain, Sweden and Switzerland.

16 people had been arrested before today's raid. One US suspect, an air force officer, later committed suicide.

Casting the net for paedophiles

The success of Operation Magenta has highlighted the need to address the increasing problem of online child pornography, writes Sarah Left

Almost every time another social tragedy comes to light now, it seems to have been internet-assisted: from selling babies like commodities (the internet twins), to fascism (Nazi memorabilia for sale on Yahoo), and, most disturbingly, paedophilia (Wonderland and Operation Magenta).

The news stories invariably lead some people to blame the internet as a haven for vice, abuse and illegal activity.

This time they may be right. The arrest of dozens of suspected paedophiles around Britain today as part of Operation Magenta highlights a wave of child pornography that is being exacerbated by the internet.

Hertfordshire Constabulary, which headed Operation Magenta, may have gone too far when it seemed to suggest that the internet could turn an otherwise law-abiding citizen into a child abuser.

Hertfordshire's Detective Inspector Keith Tilley related the story of a related operation 18 months ago, in which his force had searched the house of a man who had been abusing his 11-year-old daughter.

'He did admit that the internet started him on that path, so there's a clear danger from the web,' the detective said.

The victims of abuse in the Catholic Church can testify to the fact that the internet is not the only place where paedophiles can hide and carry on abusing. However, there seems little doubt that the internet has changed the nature of paedophilic activity, increasing victims of abuse by creating a market for images and an efficient delivery system.

Unicef estimates that 90% of investigations into paedophilia involve the internet, and the sheer numbers of images available in newsgroups and password-protected sites bears out investigation of the medium. When police around the world busted the Wonderland Club in a coordinated operation, they found 750,000 pornographic images featuring 1,236 individual children who had been victims of sexual abuse.

Online child pornographers are not simply exchanging old material they find elsewhere: a US study found that between 36% and 37% of people who exchanged paedophilic material online were also directly abusing children.

Roger Darlington, the chairman of the Internet Watch Foundation – an industry body dedicated to eliminating online child pornography – believes the internet has changed the nature of abuse. He notes that in the offline world potential paedophiles are surrounded by a massive social consensus that condemns sexual abuse of children.

'On the net you gain access to a community that legitimises your views. If you are operating in the real world, then meeting other paedophiles will require some organisation and will be difficult, but online you'll find hundreds of thousands of people who share your views worldwide.'

He feels that more children are abused to meet a growing demand for images. Wonderland, for example, demanded an 'entry fee' of 10,000 new images.

Operation Magenta focused on a sinister subculture: paedophiles using internet chatrooms to 'groom' children for abuse, gaining their trust under false identities and enticing them to meet in the real world. It is an entirely new, and internet-specific, way of ensnaring victims.

The internet has very probably increased the sexual abuse of children, both by virtue of its global reach and by promising abusers an anonymous, 'safe' environment in which to continue abusing. Hopefully the arrests in Wonderland and Magenta will cause paedophiles to doubt the safety of operating online.

Despite the enormity of the problem, child pornography represents a minute percentage of the material available online, with the IWF concerned about approximately 60 newsgroups out of 90,000 in the UK.

Other internet users who find child pornographic material online, or who have are concerned about a particular chatroom or participant, can alert the Internet Watch Foundation, who will investigate the complaint and pass it on to the police if necessary.

Have the hackers got your number?

Think your credit card details are confidential? Then think again. Highly personal information is being sold to fraudsters on the internet – and Patrick Collinson discovered just how easy it all is

By Patrick Collinson

I have no idea who Sue Darnell of Pittsburgh, Pennsylvania, is – but this week I got her MBNA credit card number, expiry date, telephone number and billing address. I've never been anywhere near Hawaii, but I've also got all the Citibank credit card details for Mr Dewitt White, who lives in Honolulu.

It took me just a few minutes to obtain this information – and a lot more besides. I simply tapped into the thriving cyber-bazaars operated over the internet where organised criminal 'carders' buy and sell your card number and your identity. The chat forums – where card details are sold for 40 cents to $5 (it's always dollars) – are usually operated out of web locations in the former USSR but the victims live predominantly in the US and western Europe.

I won't pretend to have uncovered the really big card traffickers – that's a job for the recently launched National High-Tech Crime Unit – but it is astonishing how a simple search on Google, the internet's most popular search engine, instantly routed me to message boards where fraudsters were trading credit card information.

The sites are clearly hijacked from legitimate operators and may only survive for a few days before being closed down. The biggest I found was one nominally giving weather satellite information for northern Canada. Others included a temporarily hijacked message board at an educational college in Oklahoma City. Both were impossible to access just a few days later.

One posting on the Canadian site from someone called 'Derek' said: 'We are now giving 6,000 stolen credit card numbers and the most advanced financial hacking lesson.' Even though it was almost certainly a fake offer, it had clearly excited the interest of hundreds of users who had posted follow-ups.

Elsewhere, 'Aldi', 'Aaron' and 'Mick' were trading info in a discussion entitled 'How to Steal or Hack a Valid Credit Card'.

'Joe Black' offered four credit card details and wanted to exchange more. He had found a willing buyer in someone called 'J-J'.

But these are small-scale players. According to computer security expert Dr Neil Barrett, the credit card trading centre of the world is St Petersburg in Russia. It is the site of a number of secret internet marketplaces where card details are offered in bulk, typically costing $1 a card, sold in batches of 500 through to 5,000.

'It's not done through open access websites or newsgroups. These are point-to-point chat sessions between individual groups of hackers on an IRC channel or a side channel off an IRC channel. Junior hackers move up the scale until they are invited into a senior channel with just half a dozen people exchanging tips, tricks and credit cards. Hacking a credit card is the entry requirement for people who want to move up this ladder,' says Dr Barrett.

The Russians have adopted the practices of Wall Street, with a virtual stock market in cards, where the prices rise and fall according to daily demand. The buyers tend to be from the Far East, according to Dr Barrett.

The volume of credit card fraud has now spiralled to 'horrifying' levels according to Richard Tyson Davies of APACS, the group representing the payment systems for all the major UK banks and building societies. He says that card fraud last year was £411m compared with just £135m in 1998. By 2005 he predicts that criminals will be illegally extracting £1bn a year from British cards. Britain's biggest credit card issuer, Barclaycard, lost £41m alone to credit card fraud last year.

Testament to the alarming globalisation trend in card fraud is the fact that one-third of fraudulent use of British-registered credit cards happens abroad, as 'skimmers' and hackers collect the data in the UK but use it far from the owner's home. 'Advances in technology have made it easier for organised criminal gangs to move information around the world,' says APACS.

Internet card fraud is still a relatively small, but fast-growing, part of the total sum stolen. It is centred on Britain and America (it is five times higher in Britain than in France) because that is where the largest volume of card-paid internet transactions takes place.

Hackers rarely target the databases of the banks and credit card issuers, or the transmission systems such as Visa or Mastercard. These are ferociously protected systems which to date have proved impenetrable to criminal gangs.

But there is a much easier way to obtain credit card details. The criminals target the servers of online businesses where the credit card details of customers are held. The insecurity of the internet is not in the transmission of your credit card details over the telephone line, but in how those details are then held at the other end.

Online retailers encourage a sense of security for customers by saying they offer 128-bit encryption technology that prevents your details being intercepted. This is true, and it works. But it doesn't stop a determined hacker from downloading an online retailer's entire customer database after your details have been transmitted on to it.

Dr Barrett says: 'Generally, an online shop collects the credit card details which go on to its server. This is then copied in its entirety by a hacker, who picks through for the credit card details. Only in the last few months have people cottoned on to the fact that it is the database, not the connection, where the insecurity lies. Just because something is 128-bit encrypted doesn't say anything about the server.'

But how much of this is really going on? A search through the press archives reveals few reports, in Britain at least, of online retailers who have suffered from their systems being hacked and credit card details stolen. Indeed APACS, while recognising the internet card fraud, says 'skimming' of the magnetic strips on credit cards in shops and restaurants, particularly in London, is a far bigger problem.

Dr Barrett chuckles at the suggestion that few retailers have been hacked. He works for Information Risk Management, which helps business fight hackers, and says hacking of credit card details is rife. He is currently aware of a major British online retailer which has recently had all its customers' credit card details stolen.

Dr Barrett also works for the police, and was the prosecution witness at the trial of Welsh teenager Raphael Gray. Working from his bedroom on a £700 PC, Gray hacked e-commerce sites to obtain credit card details of 25,000 internet shoppers. He even obtained the credit card details of Microsoft mogul Bill Gates and used them to have a batch of Viagra sent to his home in California.

But he escaped a jail sentence and was given a three-year rehabilitation sentence instead. After his trial, in July last year, he told reporters: 'I did the right thing in exposing how easy it is to get these credit card details. The people concerned were lucky I was not a bandit from Colombia because I could have made a fortune.'

So should you not use your credit card over the net? Dr Barrett says: 'There is still only a very small number of online merchants' databases that we can't hack through. Even the most reputable online stores can be found wanting. If you ask me would I be worried about using my credit card over the net, then the answer is yes.'

He recommends that card users should take out a separate credit card which has a low spending limit, so that if the card is stolen in cyberspace, the amount that can be taken off it is limited.

Visa, the biggest credit card payment system, is working hard to improve security at online merchants. 'Clearly this is a worldwide problem,' says Visa spokeswoman Roz Barder. 'We are very concerned that merchants store the data on customers' credit cards properly, and to this end we launched an account information security program in August last year. A lot of work is also done liaising with law enforcement agencies globally.'

APACS has set up a Fraud Intelligence Bureau, a rapid response unit that shares information between the banks and police to combat fraud.

Home secretary Jack Straw chaired the launch in April last year of a £25m initiative to tackle cybercrime. The money will go to set up a National High-Tech Crime Unit with 40 officers based in a covert location to fight a growing illicit subculture of fraud, extortion and money laundering.

At the launch it emerged that at least four large internet banks in Britain have been attacked by computer hackers, and in each case hundreds of thousands of pounds were stolen.

'When businesses say they are not being hacked they are not telling the truth,' said Bill Hughes, director general of the national crime squad.

Web of intrigue

Credit card numbers were laid bare at adult website Playboy.com in November when hackers breached its scantily clad computer security and accessed members' details. The hacker group, operating under the name 'ingreslock 1524', identified customers' names, credit card numbers and card expiry dates in an e-mail sent to each of the victims. It said it had plans to commit a $10m fraud, but later said the hacking was only designed to expose flaws in the site's security.

More than 55,000 credit card numbers were stolen in December 2000 from Creditcards.com, which processes credit card transactions for other online companies. The hackers, believed to be based in Russia, later posted the card details online when an extortion fee was not paid.

Online music retailer CD Universe lost 300,000 of its customers' credit card details to a hacker who also posted some of them on the net. The retailer refused to pay an extortion fee of $100,000 demanded in exchange for destroying the credit card files.

Western Union closed its website for five days in September last year after a security breach saw hackers access 15,000 records containing credit and debit card details.

Police in Northern Ireland last year raided a home in Belfast, described as 'looking like something out of the NASA space station at Cape Canaveral', to arrest a hacker alleged to have taken part in a £17m credit card and phone fraud. His case has yet to reach court.

Crooks set up worldwide web of deceit

From auction fraud to the Nigerian Letter scam, the Net has become a thief's paradise-reports Jonathan Lambeth

The internet goldrush brought with it the usual quota of scam artists and crafty cons, but since the stock market collapse for new economy companies, it seems it is only the crooks who are coining serious cash.

Since e-commerce first became an alternative to the high street, the fear of credit card fraud has been widely touted as a disincentive to shop online. In fact, this fear has been a red herring, with buying on the internet so far not inherently riskier than by other means.

That is not to say the internet is fiscally secure – far from it. In May last year, as part of Operation Cyber Loss, criminal charges were brought against 90 individuals and companies involved in a range of different internet or e-mail frauds in the US. More than 56,000 people lost $117m from the frauds, according to the Federal Bureau of Investigation (FBI).

Top of the least wanted in terms of internet fraud is that done using online auction websites. This includes simple misrepresentation of merchandise to non-delivery of goods. The average sum lost in the US last year in each auction fraud was $230 (£160).

According to the FBI: 'Online auction fraud has been referred to us more than any other types of [internet-related] crime.'

As a result, it set up its Internet Fraud Complaint Centre (www.ifccfbi.gov) because law enforcement officials realised there was a high volume of low-value internet frauds. The only way to catch the crooks was to centralise investigations.

Most scams are perpetrated not by single websites but by the virtual equivalent of junk mail – spam e-mail

So far in the UK, the amount of internet fraud remains low, or at least piecemeal. The police High Tech Crime Unit, created last year, has focused so far on unearthing illegal pornography rings and catching software pirates.

A spokesman for the unit said: 'We are not aware that auction fraud is a major problem in the UK yet, and we have no active operations in that area. However, who is to say that one or three years down the line certain types of crime will not become prevalent?'

Inventive criminals have found all sorts of other ways of eliciting money from unwary internet users.

Earlier this month a company in Finchley Road, London, was named by the US Federal Trade Commission (FTC) as the purveyor of a scam involving fake domain names ending '.brit' and '.usa'.

Enough people stumped up $59 a time for the company to make $1m (£700,000) before it was closed down, according to the FTC's estimates. The company's assets were frozen to allow compensation to be paid to defrauded consumers.

Most scams are perpetrated not by single websites but by the virtual equivalent of junk mail – spam e-mail. The FTC received an average 15,000 spam e-mails a day in February forwarded by unhappy consumers.

The most common frauds on the internet are not dissimilar to age-old cons in the real world, particularly the 'Nigerian Letter'.

This involves an e-mail from someone claiming to be part of the Nigerian government or a large bank. They are trying to get money out of the country (for a variety of reasons) but need some seed cash up front. In

return, they promise to share the sum 'rescued'.

Needless to say there never is a sum and victims – amazingly a lot of people fall for this tale – are asked for more and more cash. The average loss is estimated to be an astonishing $3,000 per victim, according to the FBI.

The internet is an ideal medium to send classic scams like this to large numbers of potential victims. It is also easy to generate and distribute variations on the theme.

According to FBI special agent Bob Pocica: 'The impact of the internet on white-collar crime is that there are an increased number of victims due to the speed and volume by which the internet contacts individuals. As we move forward in society with an increased reliance on using the internet for various types of transactions, the amount of fraud will increase as well.'

The great dotcom robberies

Spam scam

'Earn easy money in just a few weeks' or 'Work part time for us and retire in a year'. Deceptive junk e-mail is common but as the FTC says: 'They're intriguing. They're inviting. They're illegal.'

Many are simply chain letter suggesting you send $5 to the names at the top of the list. Others ask for a registration fee for some fabulous money-making scheme or charge for an information pack. The Nigerian Letter remains the true con man's favourite – for now.

Auction ructions

This is the easiest target for fraudsters as the buyer is entirely dependent on the seller to send the goods.

Domain name games

From offering non-existent names to trademark theft, the Arthur Daley opportunities seem never-ending.

One cyberscammer even bought 5,500 trademark or similar web addresses such as www.annakurnikova.com, to redirect unsuspecting internet users to his main website. They were then bombarded with dozens of pop-up adverts for services including porn and gambling. US authorities estimated he was earning $1m a year from the scam before he was caught.

Dodgy dialling

Redirecting an internet connection to a premium rate number is a classic means, particularly for adult services, of generating revenue. Some even route the call via countries such as Madagascar to generate outrageously high bills.

Stock option

Ramping and dumping was a popular trick when stock markets were soaring upwards. Use a website or chat area to talk up a company then sell your stock as the gullible dive in. Some websites and e-mails have even tired to get people to invest in non-existent companies before they float.

Credit unworthy

In December the American authorities shut down a Canadian operation, Consumer Resource Services, that was persuading elderly internet users to hand over their credit card details in return for the promise of free goods. It then used the card numbers to pay itself via online payment services.

Call for more government control of the net

The president of the internet's governing body yesterday made a recommendation almost unthinkable in dot.com circles: he wants to see more direct government involvement in the development of the world-wide web.

Stuart Lynn, the president of the Internet Corporation for Assigned Names and Numbers (Icann), has recommended a major restructuring, saying the goal of leaving the internet in private hands has proven unworkable.

Icann was set up in 1998 to provide independent management of the internet's development, taking over the job from the US government as the web became internationally crucial.

The body is in charge of co-ordinating the internet's addressing policies, including those for domain names such as .com and .gov.

Mr Lynn has recommended a new structure that would see governments nominating one-third of a 15-member board.

'I am now convinced that the original desire to avoid a totally governmental takeover . . . led to an overreaction – the choice of a totally private model,' he said.

He said the private model is unworkable 'because it leaves Icann isolated from realworld institutions – governments – whose backing and support are essential'.

Icann has faced questions about its legitimacy from the beginning.

Long-time internet users accuse Icann of being beholden to corporate interests, while administrators of domain names around the world have refused to recognise Icann's authority.

The proposal, which came during a weekend closed-door retreat, is likely to face significant opposition from public-interest groups, particularly for eliminating direct participation by internet users.

Under the new plan, the board would consist of 15 members: one-third nominated by governments, one-third through a committee process and the rest consisting of Icann's president and appointments by four policy and technical groups.

Michael Froomkin, a University of Miami law professor who runs an Icann watchdog site, called the proposal misguided and said it would reduce 'to even greater impotence the people who ask troubling questions'. Currently, five of the 19 board members are elected by the general internet community.

ADDITIONAL RESOURCES

You might like to contact the following organisations for further information. Due to the increasing cost of postage, many organisations cannot respond to enquiries unless they receive a stamped, addressed envelope.

British Educational Communications and Technology Agency (Becta)
Milburn Hill Road
Science Park
Coventry, CV4 7JJ
Tel: 02476 416994
Fax: 02476 411418
E-mail: becta@becta.org.uk
Web site: www.becta.org.uk
BECTA is the government agency for information and communications technology (ICT) in education. It supports and develops the use of technology to raise educational standards.

British Educational Suppliers' Association (BESA)
20 Beaufort Court
Admirals Way
London, E14 9XL
Tel: 020 7537 4997
Fax: 020 7537 4846
E-mail: besa@besanet.org.uk
Web site: www.besanet.org.uk
BESA is a trade association that draws its membership from the educational supply industry. Members include manufacturers and suppliers of educational and training equipment, material, consumables, teaching aids, furniture, technology hardware and software and services for all ranges, areas of the curriculum and administration.

Childnet International
Studio 14
Brockley Cross Business Centre
96 Endwell Road
London, SE4 2PD
Tel: 020 7639 6967
Fax: 020 7639 7027
E-mail: info@childnet-int.org
Web site: www.childnet-int.org
Childnet International was established as a non-profit organisation in 1995 with the mission to help make the Internet a great and safe place for children and to ensure that their interests are promoted and protected.

CyberAngels
Guardian Angels
Cyberangels Program
P O Box 3009, Allentown,
PA 18106, USA
E-mail: webmaster@cyberangels.org
Web site: www.cyberangels.org
CyberAngels is the largest online safety, education and help group in the world. They are a cyber-neighbourhood watch and operate worldwide in cyberspace through our more than 3,000 colunteers from more than fourteen countries.

International Labour Organization (ILO)
Millbank Tower
21-24 Mill Bank
London, SW1P 4QP
Tel: 020 7828 6401
Fax: 020 7233 5925
E-mail: london@ilo-london.org.uk
Web site: www.ilo.org/london
The International Labour Organization is the United Nations agency with global responsibility for work, employment and labour market issues.

Internet Watch Foundation (IWF)
5 Coles Lane
Oakington, Cambridge, CB4 5BA
Tel: 01223 237700
Fax: 01223 235870
E-mail: admin@iwf.org.uk
Web site: www.internetwatch.org.uk
The Internet Watch Foundation (IWF) was launched in late September 1996 by PIPEX to address the problem of illegal material on the Internet, with particular reference to child pornography.

National Family and Parenting Institute (NFPI)
430 Highgate Studios
58-79 Highgate Road
London, NW5 1TL
Tel: 020 7424 3460
Fax: 020 7424 3590
E-mail: info@nfpi.org
Web site: www.nfpi.org
An independent charity working to improve the lives of parents and families by campaigning for a more family-friendly society.

NCH
85 Highbury Park
London, N5 1UD
Tel: 020 7704 7000
Fax: 020 7226 2537
Web site: www.nch.org.uk
NCH improves the lives of Britain's most vulnerable children and young people by providing a diverse and innovative range of services for them and their families and campaigning on their behalf.

Parents' Information Network (PIN)
PO Box 16394
London, SE1 3ZP
Tel: 020 7357 9078
Fax: 020 7357 9077
E-mail: post@pin.org.uk
Web site: www.pin.org.uk
PIN is the only national independent organisation specifically dedicated to helping families get to grips with new technology – buying, using and learning with computers. Produces two unique starter packs for families, *Buying a computer and printer* (£14.99) and *Buying educational software* (£19.99), for more details on both packs and how to order call PIN on 0870 60 40 231.

The Association of Internet Hotline Providers in Europe (INHOPE)
PO Box 3010
Goring, BN12 4WF
Tel: 01903 2453793
Fax: 01903 2453793
E-mail: info@inhope.org
Web site: www.inhope.org
The INHOPE Association exists to facilitate co-operation between European Internet Hotline providers. Its mission is to eliminate child pornography from the Internet and protect young people from harmful and illegal uses of the Internet.

INDEX

ACKNOWLEDGEMENTS

The publisher is grateful for permission to reproduce the following material.

While every care has been taken to trace and acknowledge copyright, the publisher tenders its apology for any accidental infringement or where copyright has proved untraceable. The publisher would be pleased to come to a suitable arrangement in any such case with the rightful owner.

Chapter One: The Internet Society

Internet access, © Crown copyright is reproduced with the permission of the Controller of Her Majesty's Stationery Office, *Home access to the Internet*, © Crown copyright is reproduced with the permission of the Controller of Her Majesty's Stationery Office, *59% of Britons now using interactive technologies*, © MORI (Market & Opinion Research International Limited), *Online shopping in Britain increases*, © BMRB International, *Computers in schools*, © BESA (British Educational Suppliers Association), *Internet access and expansion*, © BESA (British Educational Suppliers' Association), *The net is not the place for learning*, © Telegraph Group Limited, London 2002, *Computer games 'help pupils learn'*, © Telegraph Group Limited, London 2002, *The Internet and education*, © Crown copyright is reproduced with the permission of the Controller of Her Majesty's Stationery Office, *Rural area have-nots lose out on the net*, © Guardian Newspapers Limited 2002, *UK passes 500,000 high-speed Internet connections*, © Crown copyright is reproduced with the permission of the Controller of Her Majesty's Stationery Office, *The widening digital divide*, © Guardian Newspapers Limited 2002, *Who's got mail?*, © 1996-2002 International Labour Organization (ILO), *Laugh? I nearly got the sack*, © Fran Abrams, *Post modern*, © Guardian Newspapers Limited 2002.

Chapter Two: Safety on the Internet

Knowing the dangers, © Internet Watch Foundation (IWF), *Internet safety*, © 2000-2002 National Family and Parenting Institute, *NCH's checklist for the netsmart*, © NCH, *Wise up to the net*, © Crown copyright is reproduced with the permission of the Controller of Her Majesty's Stationery Office, *Access to Internet*, © MORI (Market & Opinion Research International Limited), *Internet filtering software*, © Parents' Information Network (PIN), *Kids look here!*, © Childnet International 2002, *Overview of the problem*, © Internet Hotline Providers in Europe Association (INHOPE), *General Internet safety tips*, © CyberAngels, *The Net of corruption*, © The Daily Mail, November 2001, *Making your family Internet code*, © Parents' Information Network (PIN), *Know the rules of the cyberroad*, © familyguidebook.com, Parry Aftab, *Policing the net*, © Guardian Newspapers Limited 2002, *Police swoop on international net paedophile ring*, © Guardian Newspapers Limited 2002, *Casting the net for paedophiles*, © Guardian Newspapers Limited 2002, *Have the hackers got your number?*, © Guardian Newspapers Limited 2002, *Crooks set up worldwide web of deceit*, © Telegraph Group Limited, London 2002, *Call for more government control of the net*, © Guardian Newspapers Limited 2002.

Photographs and illustrations:

Pages 1, 8, 19, 24, 34, 37, 40: Simon Kneebone; pages 5, 20, 27: Bev Aisbett; pages 11, 13, 31: Fiona Katauskas; pages 23, 26, 32: Pumpkin House.

Craig Donnellan
Cambridge
September, 2002